Proceedings of the 6th International Workshop on Hydro Scheduling in Competitive Electricity Markets

Arild Helseth
Editor

Proceedings of the 6th International Workshop on Hydro Scheduling in Competitive Electricity Markets

 Springer

Editor
Arild Helseth
SINTEF Energy Research
Trondheim, Norway

ISBN 978-3-030-13217-0   ISBN 978-3-030-03311-8   (eBook)
https://doi.org/10.1007/978-3-030-03311-8

This Springer imprint is published by the registered company Springer Nature Switzerland AG
The registered company address is: Gewerbestrasse 11, 6330 Cham, Switzerland

Preface

The scheduling of generation resources is a key component of the electricity industry all over the world. In hydro-dominated systems, the generation scheduling problem becomes a very complex task due to the need to coordinate reservoirs under uncertainty in inflow. In a market environment, this complexity is compounded by uncertainties in electricity prices, the need for risk management, and integration with other markets—such as natural gas and carbon markets. The scheduling requires detailed modeling of system components and uncertainties in optimization and simulation models that run in reasonable computational times. A further complicating factor is that no hydro systems are alike. Each system is uniquely defined, e.g., by watercourse topology, man-made storages, release elements used to control the water flows, constraints imposed on the operation, and the regulatory framework governing the operation.

Either in a cost-based or profit-maximization framework, the coordination of the operation of a hydro system implies application of computer models and tools. The overall scheduling problem is normally divided into a hierarchy of scheduling problems with different planning horizons and degrees of detail in the representation of the system and of the related uncertainties. Different methodologies are utilized, including stochastic dynamic programming, decomposition-based methods, linear and nonlinear programming, and heuristics. Results from the long-term scheduling models are to be used as input to more detailed mid-term models, which in turn feed their results into the short-term scheduling procedures. The computational challenge is usually overcome by means of parallel processing and the use of computing clusters.

Looking into the future, the field of hydro scheduling faces many important challenges that need innovative and functional solutions. In this context, the International Workshop on Hydro Scheduling in Competitive Electricity Markets has emerged as an intimate and worldwide forum for researchers and practitioners to present the latest research results, ongoing developments, best practices and applications related to hydro scheduling.

Arild Helseth

Organization

The International Workshop on Hydro Scheduling in Competitive Markets was first organized in Trondheim, Norway, in 2002. Other successful editions were held in Stavanger (2005), Oslo (2008), Bergen (2012), and Trondheim (2015).

This volume presents selected papers from the 6th International Workshop on Hydro Scheduling in Competitive Electricity Markets, arranged September 12–13, 2018, in Stavanger, Norway. This workshop was organized by SINTEF Energy Research and the Norwegian University of Science and Technology, and the Organizing Committee comprised the following members:

Michael Belsnes, SINTEF Energy Research, Norway
Birger Mo, SINTEF Energy Research, Norway
Eline Opdalshei, SINTEF Energy Research, Norway
Arild Helseth, SINTEF Energy Research, Norway
Martin N. Hjelmeland, Norwegian University of Science and Technology, Norway

All workshop presenters were encouraged to prepare a full scientific article. After a careful peer review process, a total of ten papers were selected to be included in this volume. The articles address the thematic areas such as (a) computational and methodological advances in hydro scheduling, (b) renewable integration and hydro scheduling, (c) hydro scheduling in multiple power markets, and (d) practical experiences and best practices in hydro scheduling.

All articles presented in this volume have gone through a peer review process organized by Arild Helseth and the members of the Scientific Committee:

André Luiz Diniz, CEPEL, Brazil
Hubert Abgottspon, HES-SO Valais-Wallis, Switzerland
Juan Ignacio Pérez-Díaz, Universidad Politécnica de Madrid, Spain
Luiz Augusto Barroso, Energy Research Company, Brazil
Magnus Korpås, Norwegian University of Science and Technology, Norway

Nils Löhndorf, University of Luxembourg, Luxembourg
Olav Bjarte Fosso, Norwegian University of Science and Technology, Norway
Pascal Côté, Rio Tinto Alcan, Canada
Stein-Erik Fleten, Norwegian University of Science and Technology, Norway

Contents

Blackbox Optimization for Chance Constrained Hydro Scheduling Problems

Sara Séguin[1(✉)] and Pascal Côté[2]

[1] Université du Québec à Chicoutimi, Saguenay, QC G7H 2B1, Canada
`sara.seguin@uqac.ca`
[2] Rio Tinto, Power operation, Saguenay, QC G7S 4R5, Canada
`pascal.cote@riotinto.com`

Abstract. This paper presents a novel method to treat a chance constrained formulation of the hydropower reservoir management problem. An advantage of this methodology is that it is easily understandable by the decision makers. However, when using explicit optimization methods, the optimal operating policy requires to be simulated over multiple scenarios to validate the feasibility of the constraints. A blackbox optimization framework is used to determine the parameters of the chance constraints, embedding the chance constrained optimization problem and the simulation as the blackbox. Numerical results are conducted on the Kemano hydropower system in Canada.

Keywords: Hydropower reservoir management
Blackbox optimization · Chance constrained optimization
Stochastic dynamic programming

1 Introduction

Recent years have shown variability in meteorological and hydrological forecasts. Whereas not so long ago, years seemed to repeat themselves and follow a certain trend, managing efficiently hydropower systems has become increasingly difficult for many reasons, especially due to the variability of the inflows. Hydropower is a clean and renewable energy and in the province of British-Columbia, 90% of the energy is provided by hydropower. It is in everyone's interest to produce the most energy out of the available water. The management of power plants and turbines is not only difficult due to the stochastic inflows, but also given the nonlinearities that exist in the mathematical formulations [1] of such problems. Power produced by a turbine is a nonlinear function of the unit water flow and the net water head, which is a function of the total water flow at the plant, which also affects the tailrace elevation. Modeling these functions presents a challenge, since their approximation have an impact on the solutions obtained from the optimization models. Other constraints increase the difficulty of these problems, such as bounds on reservoir levels for dam safety or leisure activities such as

© Springer Nature Switzerland AG 2019
A. Helseth (Ed.): HSCM 2018, *Proceedings of the 6th International Workshop on Hydro Scheduling in Competitive Electricity Markets*, pp. 1–7, 2019.
https://doi.org/10.1007/978-3-030-03311-8_1

beaches and navigation requirements. Water flow constraints for environmental protection and flood control also need to be considered, as well as energy production requirements [2]. The reservoir management problem [3] consists in determining the reservoir levels and water flows at the power plants given a time horizon, usually weekly decisions on a yearly horizon. Depending on the characteristics of the system and the random processes involved, it may be impossible to satisfy all of the constraints presented above. Therefore, the optimization models should account for multiple criterias when seeking a trade-off solution, thus solving a multiobjective optimization problem. The uncertainty of the inflows [4] prevents the use of multiobjective optimization since a decision has to be made before the realization of the uncertainty. Solving the problem in a multiobjective context would actually add to the complexity of the problem, leading to increased difficulty in the decision-making process. In this case, probabilistic constraints [5,6] are an interesting avenue. Probabilistic constraints allow the constraints to be violated, given a certain probability. From a decision-making point of view, it could be acceptable that the reservoir bounds may be violated a certain number of times during the year, for example. In practice, a penalty term is added to the objective function to account for a constraint violation. At first view, this method is easy to implement, but as the reservoir management problem is stochastic, many scenarios are used when solving the optimization problem, therefore the penalty needs to be adjusted to consider many scenarios. Also, the more probabilistic constraints, the more penalties are there to adjust. Parameters adjustments is usually neglected, although they may have a significant influence on the quality of the solution. Often, experience of the engineers is taken into account but there is no real measure of good parameters. In this paper, we propose a novel approach to adjust automatically the parameters of the penalties, while solving the reservoir optimization problem. A blackbox optimization [7] solver is used to optimize the values of the penalties associated to the probabilistic constraints. The reservoir management problem is solved concurrently, leading to an automatic adjustment of the penalties. Recently, the authors have used blackbox optimization [8] to find the best scenario tree parameters to represent the inflows in multi-stage stochastic short-term optimization problem, which is a promising avenue for this study. The paper is organized as follows. Section 2 present the hydropower system studied in this paper. The reservoir management problem is exposed in Sect. 3. Section 4 details the formulation of the blackbox optimization problem for the automatic adjustment of the parameters. Numerical results are available in Sect. 5 and final remarks are discussed in Sect. 6.

2   Case Study

The case studied in this paper is the Kemano hydropower system, owned and operated by Rio Tinto to feed the aluminium smelters located in Kitimat. It is situated in northern British-Columbia, Canada and includes a reservoir that releases water to a powerhouse through a 10 Km tunnel to the Pacific ocean. The spilling water is released on a land near the Nechako river. Figure 1 illustrates the Kemano hydropower system. Reservoir storages are given by s_1, s_2, inflows by q_1, q_2, q_3, water processed u_1 and v_1 water spilled at Kemano power plant, and outflow rated at Cheslatta Lake by v_2, v_3.

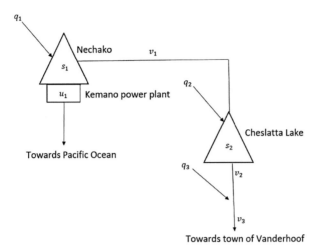

Fig. 1. Kemano hydropower system

3   Reservoir Management Problem

The operation of the reservoir consists in finding the best water releases policy that maximizes the energy production while respecting several operational constraints. The problem is formulated as a stochastic dynamic programming algorithm and is explicitly presented in [9]. In brief, the problem is multiobjective by its nature. The reservoir management policy must seek a trade-off solution between flooding the town of Vanderhoof and supplying the required energy at the aluminium smelter. In this paper, we propose a formulation of the reservoir management problem using chance constraints (CC), leading to a single objective problem.

3.1   Chance Constraints

For the sake of clarity, the following problem presents only the chance constraints. Therefore, usual and necessary water balance constraints, and bounds on water flows and reservoirs are dropped. The objective is to maximize the energy production while respecting the chance constraints:

$$\max_{u_t} \mathbb{E}\left[\sum_{t=1}^{T} P_t(s_t, u_t, v_t, q_t)\right] \tag{1}$$

subject to

$$\Pr\left(v_{1,t} < v_{1,t}^{min}\right) \leq \xi_{1,t}, \ \forall t \in 1, 2, \ldots, T, \tag{2}$$

$$\Pr\left(v_{1,t} > v_{1,t}^{max}\right) \leq \xi_{2,t}, \ \forall t \in 1, 2, \ldots, T, \tag{3}$$

$$\Pr\left(P_t < P_t^{min}\right) \leq \xi_{3,t}, \ \forall t \in 1, 2, \ldots, T. \tag{4}$$

where $P_t(\cdot)$ is the energy production function, P_t^{min} is the minimum energy production (Eq. (2)), $v_{1,t}^{max}$ is the maximum flow to avoid downstream flooding (Eq. (3)), $v_{1,t}^{min}$ is the minimum environmental flow (Eq. (4)), ξ are the probabilities associated to respecting a constraint and T is the total number of periods. Chance constraints are used because even with a perfect foresight of natural inflows, it is impossible to respect all of the constraints, leading to an infeasible problem. To deal with these infeasibilities, the problem is formulated with chance constraints. The three above mentioned constraints are formulated to meet a certain level, restricting the feasible region to have a high confidence level. In practice, penalty parameters χ_1, χ_2, χ_3 are added to the objective function to account for the chance constraints violations and yields the following optimization model:

$$\max_{u_t} \mathbb{E}\left[\sum_{t=1}^{T} P_t(s_t, u_t, v_t, q_t) - \chi_1(v_{1,t}^{min} - v_{1,t}) - \right. \tag{5}$$

$$\left. \chi_2(v_{1,t} - v_{1,t}^{max}) - \chi_3(P_t^{min} - P_t)\right]$$

subject to

$$Eq.(2) - (4). \tag{6}$$

Since the policy obtained from the above problem (Eq. (5)–(6)) is simulated over different scenarios, parameters are proper to a scenario. Therefore, a set of parameters that fit all of the scenarios need to be found in order to find an optimal policy when simulated over a large set of scenarios.

4   Blackbox Optimization

A blackbox (BB) optimization solver is used to optimize the values of the penalties associated to the chance constraints leading to an automatic adjustment of the penalties. BB optimization is used when the objective function and/or the constraints do not have an analytical representation. In this case, the reservoir management problem is modeled as the BB, i.e. the operating policy is derived by a SDP with specific penalty parameters and simulated over a subset of inflow scenarios called the calibration set. The SDP problem is implicitly solved with the values of χ provided by the solver. The objective function of the BB optimization problem is given by Eq. (5) and the decision variables are respectively χ_1, χ_2 and χ_3. The solver used in the paper is NOMAD [10], an implementation of the Mesh Adaptive Direct Search Methods (MADS) [11]. This methods seeks for a sequence of iterates that improves the value of the objective function given a set of directions that lie on a mesh. The mesh is refined or coarsened given the new iterate improves or deteriorates the current solution. The method iterates until convergence to optimality or given a budget of evaluations specified by the user. In this paper, we first use a BB optimization framework to find the best values for the penalty parameters associated to violating the CC, then use these values of parameters to simulate the operating policy. The process is shown in Fig. 2. The calibration part of the process consists in finding the best values of χ_1, χ_2 and χ_3 using NOMAD. A calibration set of scenarios is used for the SDP reservoir management problem. The simulation part of the process consists in simulating the policy obtained from the calibration process with a validation set of scenarios and the best parameters for χ_1, χ_2 and χ_3 to evaluate the robustness of the optimal penalty parameters. The current criteria used to evaluate robustness is the violations of the CC in the simulation phase.

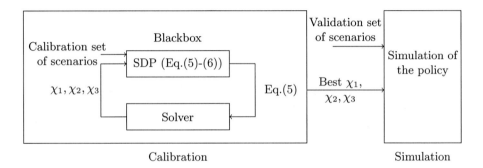

Fig. 2. Calibration and simulation process

5 Results

Five test runs have been conducted and results are shown in Table 1. Each of these test runs have a different starting point, more precisely a set of penalty parameters. The objective function is translated into the mean over the five runs of the yearly average relative production increase[1], in order to be understandable by the decision makers, and is visible in the second column. Note that the real values of the objective function are not reported since they are difficult to interpret, but rather recomputed to a relative production increase with the policies obtained. The best penalization parameters obtained from the BB solver on the calibration and validation sets are used when simulating the SDP operating policy. The third column shows the number of times one of the three CC is violated, respectively (Eqs. (2), (3), (4)), and possible values range from 0 to 5, the number of test runs. For the calibration set of scenarios, 500 scenarios are generated synthetically and for the validation set of scenario, 63 historical

---

[1] Due to confidentiality reasons, the revenue values have been scaled and do not represent the real revenue.

sequences are available. Future work based on the choice of the scenarios and the generation of synthetic scenarios will be carried out. One conclusion that it is possible to draw at this time is that the BB optimization framework seems to provide inconsistent results based on CC violation when comparing calibration and validation results. Figure 3 presents the value of the objective function versus the iterations of the BB solver, for one test run on the calibration set. A circled value indicates the solution is feasible, thus does not violate the CC. The BB solver succeeds to improve the objective function and finds a feasible solution on the calibration set. On the down side, the optimal parameters with the validation set of scenarios violates one of the CC. As this is on-going research, extensive tests will be carried out in a near future to improve the robustness of the optimal parameters obtained from the calibration set of scenarios.

Table 1. Calibration and validation results of 5 blackbox optimization runs

Subset	Yearly average relative production increase	Total number of CC violated	Number of scenarios	Average computation time in sec
Calibration	0%	(0, 0, 0)	500	15482
Validation	0.73%	(0, 0, 5)	63	< 1

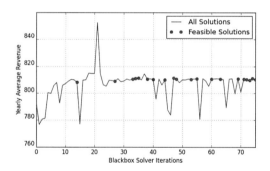

Fig. 3. Objective function versus BB iterations, for one test run on the calibration set of scenarios

6   Conclusion

This paper uses a blackbox optimization framework to treat a chance constrained formulation of the hydropower reservoir management problem solved with a stochastic dynamic programming algorithm. Penalties are added to the objective function to permit constraints violations and optimal penalty values are found by formulating the reservoir management problem as a blackbox optimization problem. A calibrating set of inflow scenarios is used to find the penalties and a validation set of inflow scenarios is used to assess the quality of the solution. Numerical results are preliminary but promising and extensive tests will be carried out in a near future. The number of scenarios used in the calibration and validation sets will be investigated, to validate the effect on the solution to the reservoir management problem.

References

1. Finardi, E.C., Scuzziato, M.R.: Hydro unit commitment and loading problem for day-ahead operation planning problem. Int. J. Electr. Power Energy Syst. **44**(1), 7–16 (2013)
2. Pérez-Díaz, J.I., Wilhelmi, J.R., Arévalo, L.A.: Optimal short-term operation schedule of a hydropower plant in a competitive electricity market. Energy Convers. Manag. **51**(12), 2955–2966 (2010)
3. Labadie, J.W.: Optimal operation of multireservoir systems: state-of-the-art review. J. Water Resour. Plan. Manag. **130**(2), 93–111 (2004)
4. Eum, H.-I., Kim, Y.-O.: The value of updating ensemble streamflow prediction in reservoir operations. Hydrological Processes **24**(20), 2888–2899 (2010)
5. Andrieu, L., Henrion, R., Römisch, W.: A model for dynamic chance constraints in hydro power reservoir management. Eur. J. Oper. Res. **207**(2), 579–589 (2010)
6. Turgeon, A.: Daily operation of reservoir subject to yearly probabilistic constraints. J. Water Resour. Plan. Manag. **131**(5), 342–350 (2005)
7. Audet, C.: A survey on direct search methods for blackbox optimization and their applications. In: Pardalos, M.P., Rassias, M.T. (eds.) Mathematics Without Boundaries: Surveys in Interdisciplinary Research, pp. 31–56. Springer, New York (2014)
8. Séguin, S., Audet, C., Côté, P.: Scenario-tree modeling for stochastic short-term hydropower operations planning. J. Water Resour. Plan. Manag. **143**(12), 04017073 (2017)
9. Desreumaux, Q., Côté, P., Leconte, R.: Comparing model-based and model-free streamflow simulation approaches to improve hydropower reservoir operations. J. Water Resour. Plan. Manag. **144**(3), 05018002 (2018)
10. Le Digabel, S.: Algorithm 909: NOMAD: nonlinear optimization with the MADS algorithm. ACM Trans. Math. Softw. **37**(4), 44:1–44:15 (2011)
11. Audet, C., Dennis Jr., J.E.: Mesh adaptive direct search algorithms for constrained optimization. SIAM J. Optim. **17**(1), 188–217 (2006)

Evaluating Approaches for Estimating the Water Value of a Hydropower Plant in the Day-Ahead Electricity Market

Ignacio Guisández$^{(\boxtimes)}$ and Juan Ignacio Pérez-Díaz

Universidad Politécnica de Madrid, Calle Profesor Aranguren 3,
28040 Madrid, Spain
i.guisandez@upm.es

Abstract. This paper addresses the question of whether the use of complex algorithms, based on mixed integer linear programming, to solve the intrastage decision problems of a stochastic dynamic programming (SDP) based annual scheduling model aimed to calculate the water value of a hydropower plant is a fruitful effort. To this purpose, four 1000-year long simulations using the water value obtained from four different optimisation SDP-based scheduling models (three using mixed integer linear programming to solve the intrastage decision problems and other using linear programming) are compared. The results suggest that the small increase in profit does not make up for the necessary increase in computational time. Nonetheless, the study should be replicated using other hydropower plants and more complicated topologies in order to get more sound conclusions.

Keywords: Linear programming · Mixed integer linear programming
Stochastic dynamic programming · Water value

1 Introduction

The concept of *watervalue* (WV) in the hydroelectric field has been defined in different ways in the literature. Among them, one suitable for deregulated electricity markets, is that one that describes WV as the marginal change in the hydropower producer's expected profit for a marginal change in its available hydro resources [1]. Given its important role as a bridge between the long- and short-term hydro scheduling [2] and the current difficulty in its determination [3], the analysis of the aspects involved in WV calculation is still interesting. This study is focused on how to model two of those aspects: the hydro unit start-ups and the *generation characteristic* of the plant. A priori, it is reasonable to assume that the higher the level of detail of each of those aspects incorporated in a WV calculator, the greater the profit obtained through the use of the resulting

© Springer Nature Switzerland AG 2019
A. Helseth (Ed.): HSCM 2018, *Proceedings of the 6th International Workshop on Hydro Scheduling in Competitive Electricity Markets*, pp. 8–15, 2019.
https://doi.org/10.1007/978-3-030-03311-8_2

WV. However, some experiences related to hydro scheduling have shown that this is not always the case [4].

The objective of the current research is to delve into the importance of considering the integrality of hydro unit start-up variables as well as to compare several methods for modelling the plant generation characteristic for the WV calculation. In order to facilitate the exposition and discussion of the results, a representative theoretical 1-unit hydropower plant associated to a single reservoir that hypothetically participates in the Spanish day-ahead electricity market is analysed as a case study. The idea is first to calculate the plant WV in different ways, and then simulate the plant long-term operation using each of these WV with the aim of comparing the resulting profit.

Four different optimisation models for WV calculation were developed. Using the same *stochastic dynamic programming* (SDP) based approach proposed in [5], the models differ in the formulation of the intrastage decision problems. Thus, three of the models use a formulation based on *mixed integer linear programming* (MILP) whereas the other on *linear programming* (LP). The three models with MILP differ from each other in the method used for modeling the generation characteristic: one follows [6], another [7], and the other [8]. The model with LP follows [9] and [10]. In addition, a short-term scheduling model, based on MILP, was developed and used to simulate the plant long-term operation.

It is important to note that this study has some similarities with [11]. Both the MILP formulation proposed in [8] and a relaxed version of that formulation were used in [11] to calculate the WV. Then, several simulations were run using the WV obtained with the two formulations. Both the MILP and the LP formulations were used to perform the simulations. This paper takes a step ahead of [11] by using three different MILP to model the generation characteristic of the plant, the one used in [11], and the two most highly cited ones, the ones proposed in [6] and [7]. The LP formulation used in this paper to model the generation characteristic in one of the WV calculators, is not a relaxed version of any of the MILP ones, but rather is based on the most highly cited linear formulation for such a purpose [9].

The paper is organised in the following manner. In Sects. 2 and 3, the optimisation models and the case study are qualitatively described. The main results are given in Sect. 4. Finally, in Sect. 5 are the conclusion and future proposals.

2   Optimisation Models

2.1   WV Calculators

The four WV calculators used in the study have the same SDP algorithmic structure presented in [5] which can be summarised as follows. It has an annual planning period with weekly decision stages. The state variables are: water volume stored in the reservoir at the beginning of the week, water inflow volume of the week, and average energy price of the week. The first variable is discretised in nine equidistant values [12]. The two latter variables are modelled each by means of a Markov chain [13].

The optimal profits corresponding to the weekly decision problems, with hourly time steps, involved in the above-outlined state space are found either by MILP (three of the WV calculators) or by LP (the remaining one). Regardless of the approach used, the equations considered are the same except those that concern the plant generation characteristic (and that will be described in the following subsubsections). Both the water inflow volume and the average energy price of the week are assumed known in each weekly decision problem, i.e. one weekly decision problem is solved for each realization of the said pair of variables. Obviously, all variables in the LP formulation are continuous, whereas in the MILP formulations some of them are integer. Finally, another distinguishing feature of the WV calculators using MILP formulations is a recalculation, with the real plant generation characteristic (i.e., not linearised), of the plant hourly power outputs according to the decisions provided by the MILP algorithm in order to obtain more precisely the weekly profit. The recalculation is not performed by the WV calculator using LP since it provides in some cases infeasible values of the plant hourly power outputs. A heuristic criterion would therefore be necessary for the WV calculator using LP to perform such a recalculation.

The objective function is common to the four WV calculators and consists in maximising the profit obtained from selling energy in the day-ahead electricity market taking into account not only the water value at the end of the week (through a piecewise concave linear curve), but also the costs caused by the start-ups of the hydro unit as well as its wear and tear costs due to variations in the generated power. A more detailed description of the profit-to-go function can be found in [5]. That function is subject to the following constraints: the water balance (including the evaporation and seepage losses); the maximum and minimum stored volume capacities of the reservoir; the maximum flows of the hydro unit (according to the stored volume), bottom outlets, and spillway; and the decrement in the generation due to the tailrace elevation caused by the water released through the bottom outlets and the spillway.

A brief description of the formulation used to solve the weekly decision problems in the four WV calculators is included below. Hereinafter the MILP formulations will be referred to as MILP-1/2/3, and the LP one will be referred to as LP.

MILP-1: Following the method proposed in [6], the generation characteristic is modeled by a piecewise linear concave power-discharge function with breakpoints at the points corresponding to minimum flow, best efficiency and maximum flow. The curve used in each weekly decision problem corresponds to the water volume stored in the reservoir at the beginning of the week. A binary variable is used to model the hydro unit start-ups.

MILP-2: Following now [7], three power-discharge piecewise linear curves are used in this case. Unlike [7], the curves are concave. The breakpoints of the curves are selected with the same criterion as in the MILP-1 formulation. The curves used in each weekly decision problem correspond each to a different head.

The selected heads are uniformly distributed over the feasible head range of each weekly decision problem. A binary variable is used to model the hydro unit start-ups and two more to select each hour one power-discharge curve as a function of the actual head.

MILP-3: The formulation is based on the one-dimensional method coined by [8]. The generation characteristic is modeled by the same three piecewise linear power-discharge curves as in the MILP-2 formulation. As in the previous MILP formulations, a binary variable is used to model the hydro unit start-ups. It is important to note that the MILP-3 formulation uses one more binary variable than the MILP-2.

LP: The formulation is based both on [9] and [10]. The generation characteristic is modeled by a piecewise linear concave power-discharge curve with the same breakpoints as in the MILP-1 formulation.

2.2   Simulator

The formulation of the optimisation model used to simulate the short-term generation scheduling of the plant is identical to the MILP-3 formulation. This choice is supported by the results offered by [14] which showed the efficiency of this method in the short-term scheduling.

3   Case Study

One theoretical hydropower plant, designed from the average data extracted from [15], is used as a case study with the aim of easing the exposition and discussion of the results. It has 55 MW of installed power capacity, a single Francis hydro unit, and 65 m of maximum gross head; the performance curves of the unit, taken from [16], have been suitably adjusted to these data. One percent of the rated head is assumed in the generation characteristic as total (conduits and turbine) hydraulic losses [16] and a curve similar to the one proposed in [17] is used to model the influence of the flow released through the bottom outlets and the spillway on the tailrace water level.

The plant is hypothetically located in the Northwest of Spain in such a way that it participates in the Spanish day-ahead electricity market [19], receives water inflows according to a distribution pattern corresponding to Spanish oceanic fluvial data extracted from a real gauging station [18], and is subjected to evaporation rates congruent with this location [20, 21]. The orders of the Markov chains, used to model the stochasticity of the water inflow volume of the week and of the average energy price of the week (both one), have been selected through the *Akaike information criterion* and the number of classes per chain (both three) have been determined according to the length of the available historical series of these variables (51 years of daily values for the water inflow and 15 years of hourly values for the energy price). Both the head-storage curve and the surface-storage curve have been estimated following [22] and properly fixed at the Spanish average storage-power ratio estimated from [23,24]. Hence, the reservoir has a storage capacity of $256.09 \, Mm^3$. Moreover, $5.7077 \cdot 10^{-4}\%$ of the stored water volume is assumed as hourly seepage losses [25].

Finally, from the above-mentioned Markov chains, a synthetic 1000-year long series, in which both variables have hourly discretisation, was generated for the simulations. The disaggregations of these two variables from the weekly values considered in the Markov chains to hourly ones were performed by weighting the historical average weekly profile of each variable in every week by a ratio between the considered value in its respective chain and the mean value of the said average profile [5]. Figure 1 depicts the average, maximum, and minimum hourly values of the synthetic weekly profiles of the energy prices used in the simulations.

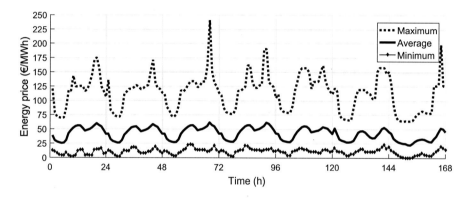

Fig. 1. Average, maximum and minimum hourly values of the synthetic weekly profiles of the energy prices.

All the models were coded in MATLAB® and GAMS^{TM} (using CPLEX^{TM}) and the simulations were carried out by a core of an Intel® Xeon® E5 at 3.1 GHz and 64 GB RAM.

4   Results

Table 1 shows the average annual profit, the average head, and the time involved in the WV calculation corresponding to each model obtained in the simulations; these results have been sorted in descending order of profit. Despite the high volatility of the simulated electricity market (Fig. 1), it can be seen that all the WV calculators using MILP formulations provide only slightly better profit than the one using a LP formulation. This is because the decisions are practically independent of the used WV, as evidenced by the close similarity of the average head resulted in the simulations. Finally, as expected, the computation times spent by the WV calculator using MILP formulations in calculating the WV are significantly greater than the one by the model using a LP formulation. The results may well be influenced by the discretisation of the state variables used for the WV calculation. However, as deduced from [4], the conclusions would most likely remain the same with a finer discretisation.

Table 1. Main results of the simulations.

WV calculator	Average annual profit		Average head		WV calculation time	
	[M€]	Variation	[m]	Variation	[h]	Variation
MILP-1	7.5901		58.56		1.9	
MILP-2	7.5896	−0.01%	58.68	0.21%	62.6	3,249.44%
MILP-3	7.5881	−0.03%	58.62	0.11%	4.6	145.13%
LP	7.5803	−0.13%	58.94	0.66%	1.6	−14.79%

5 Conclusion and Future Proposals

This study poses the question whether the use of highly developed algorithms, based on mixed integer linear programming, to solve the intrastage decision problems of a stochastic dynamic programming based medium-term scheduling model aimed to calculate the water value of a hydropower plant is an appropriate endeavour. The results reported here seem to indicate that the answer is no. The marginal improvement obtained does not seem to justify the additional complexity introduced into the procedure, especially if one considers the tremendous increase in the computational time in a case with a single hydropower plant equipped with a single unit.

Our future research directions will be to replicate this study using other hydropower plants and topologies, equipped with different numbers and types of units, in order to validate the preliminary conclusions obtained in this paper, and to study the convenience of calculating the water value considering the influence of environmental constraints. In this latter case, using integer variables to model the hydro unit start-ups might be more relevant than in the case analysed in this paper.

References

1. Reneses, J., Barquín, J., García-González, J., Centeno, E.: Water value in electricity markets. Int. Trans. Electr. Energ. Syst. **26**(3), 655–670 (2015)
2. Fosso, O.B., Belsnes, M.M.: Short-term hydro scheduling in a liberalized power system. In: 4th International Conference on Power System Technology. IEEE, Singapore, 21–24 November 2004

3. Gjelsvik, A., Mo, B., Haugstad, A.: Long- and medium-term operations planning and stochastic modelling in hydro-dominated power systems based on stochastic dual dynamic programming. In: Handbook of Power Systems I, pp. 33–55. Springer, Berlin (2010)
4. Bogardi, J.J., Budhakooncharoen, S., Shrestha, D.L., Nandalal, K.D.W.: Effect of state space and inflow discretization on stochastic dynamic programming based reservoir operation rules and system performance. In: 6th Congress Asian and Pacific Regional Division. IAHR, Kyoto, 20–22 July 1988
5. Guisández, I., Pérez-Díaz, J.I., Wilhelmi, J.R.: The influence of environmental constraints on the water value. Energies **9**(6), 446 (2016)
6. Chang, G.W., Aganagic, M., Waight, J.G., Medina, J., Burton, T., Reeves, S., Christoforidis, M.: Experiences with mixed integer linear programming based approaches on short-term hydro scheduling. IEEE Trans. Power Syst. **16**(4), 743–749 (2001)
7. Conejo, A.J., Arroyo, J.M., Contreras, J., Villamor, F.A.: Self-scheduling of a hydro producer in a pool-based electricity market. IEEE Trans. Power Syst. **17**(4), 1265–1272 (2002)
8. D'Ambrosio, C., Lodi, A., Martello, S.: Piecewise linear approximation of functions of two variables in MILP models. Oper. Res. Lett. **38**(1), 39–46 (2010)
9. Piekutowski, M.R., Litwinowicz, T., Frowd, R.: Optimal short-term scheduling for a large-scale cascaded hydro system. In: Power Industry Computer Application Conference. IEEE, Phoenix, 4–7 May 1993
10. Warland, G., Haugstad, A., Huse, E.S.: Including thermal unit start-up costs in a long-term hydro-thermal scheduling model. In: 16th Power Systems Computation Conference. PSCC, Glasgow, 14–18 July 2008
11. Helseth, A., Fodstad, M., Askeland, M., Mo, B., Nilsen, O.B., Pérez-Díaz, J.I., Chazarra, M., Guisández, I.: Assessing hydropower operational profitability considering energy and reserve markets. IET Renew. Power Gen. **11**(13), 1640–1647 (2017)
12. Goulter, I.C., Tai, F.K.: Practical implications in the use of stochastic dynamic programming for reservoir operation. JAWRA **21**(1), 65–74 (1985)
13. Gjelsvik, A., Belsnes, M.M., Haugstad, A.: An algorithm for stochastic medium-term hydrothermal scheduling under spot price uncertainty. In: PSCC 1999, Trondheim, Norway, 28 June–2 July 1999
14. Borghetti, A., D'Ambrosio, C., Lodi, A., Martello, S.: An MILP approach for short-term hydro scheduling and unit commitment with head-dependent reservoir. IEEE Trans. Power Syst. **23**(3), 1115–1124 (2008)
15. IFC: Hydroelectric power. A guide for developers and investors (2015). http://www.ifc.org/wps/wcm/connect/topics_ext_content/ifc_external_corporate_site/sustainability-at-ifc/publications/hydroelectric_power_a_guide_for_developers_and_investors
16. Krueger, R.E., Winter, I.A., Walters, R.N., Bates, C.G.: Selecting hydraulic reaction turbines. A water resources technical publication engineering monograph, no. 20. Technical report, USBR, Denver, USA (1976)
17. El-Hawary, M.E., Christensen, G.S.: Optimal Economic Operation of Electric Power Systems, vol. 26. Academic Press, New York (1979)
18. OMIE: Spanish day-ahead market data (2018). http://www.omel.es/files/flash/ResultadosMercado.swf
19. CEDEX: Spanish water inflow data (2018). http://ceh-flumen64.cedex.es/anuarioaforos/default.asp

20. Dragoni, W., Valigi, D.: Contributo alla stima dell'evaporazione dalle superfici liquide nell Italia Centrale. Geologica Romana **30**, 151–158 (1994)
21. INE: Spanish temperature data (1972). http://www.ine.es/inebaseweb/pdfDispacher.do;jsessionid=09739D168C9F3EF056383E89BACE7A07. inebaseweb01?td=183760
22. Lehner, B., Liermann, C.R., Revenga, C., Vörösmarty, C., Fekete, B., Crouzet, P., Döll, P., Endejan, M., Frenken, K., Magome, J., Nilsson, C., Robertson, J.C., Rödel, R., Sindorf, N., Wisser, D.: High-resolution mapping of the world's reservoirs and dams for sustainable river-flow management. Front. Ecol. Environ. **9**(9), 494–502 (2011)
23. ENDESA: Endesa. Centrales hidráulicas en España (2004)
24. IBERDROLA: Grandes presas (2006)
25. Gleick, P.H.: Environmental consequences of hydroelectric development: the role of facility size and type. Energy **17**(8), 735–747 (1992)

Coordinated Hydropower Bidding in the Day-Ahead and Balancing Market

Ellen Krohn Aasgård[(✉)]

Norwegian University of Science and Technology,
Alfred Getz vei 3, 7491 Trondheim, Norway
Ellen.Aasgard@ntnu.no

Abstract. Power producers with flexible production systems such as hydropower may sell their output in the day–ahead and balancing power markets. We present how the coordination of trades across multiple markets may be described as a stochastic program. Focus is on how the information structure inherent in the multi–market setting is represented through the scenario tree and mathematical modelling. In the model, each market is represented by a price or premium and an upper limit on the volume that can be traded at the given price. We illustrate our modelling by comparing coordinated versus sequential bidding strategies.

Keywords: Electricity markets · Hydropower
Stochastic programming · Scenario generation

1 Introduction

Most European power markets are organized as day–ahead auctions where expected production and consumption for the next day is traded. However, due to unforeseen events that may happen between closure of the day–ahead market and real–time, the transmission system operator (TSO) is responsible for maintaining the instantaneous balance between supply and demand. To accomplish this, the TSO procures several types of reserves from the agents in the power system. Usually, reserve products are defined based on response time, and referred to as frequency containment (primary), frequency restoration (secondary) and replacement (tertiary) reserves. In this paper, the term balancing market will be used to describe the market where the TSO procures replacement reserves. The TSO is the only buyer in the balancing market and the supply side are producers or consumers with flexible portfolios. To participate in the balancing market, agents must be able to ramp up or down a given minimum amount in a short time interval. Approved participants submit their willingness to ramp up/down, and the TSO chooses the most cost–efficient bids as need arises.

Hydropower is well suited for participating in the balancing market because of low start–up cost and the possibility of storing water in reservoirs.

© Springer Nature Switzerland AG 2019
A. Helseth (Ed.): HSCM 2018, *Proceedings of the 6th International Workshop on Hydro Scheduling in Competitive Electricity Markets*, pp. 16–25, 2019.
https://doi.org/10.1007/978-3-030-03311-8_3

The question raised in this paper is how a flexible hydropower producer may maximize its revenues from participating in both the day–ahead and balancing market. Several works have investigated optimization models for multi–market trade of electricity, see for instance [1] that uses stochastic programming [2]. The reason why stochastic optimization is appropriate, is that the trading strategy must be determined prior to market clearing, i.e. when prices are still unknown.

In this work, we present a stochastic program for a hydropower producer that coordinates its trades between the day–ahead and the balancing market. We focus on the information structure inherent in the multi–market setting and how this is represented through the scenario tree and mathematical modelling. The optimization model is an extension of [4] which showed how optimal bids for the day–ahead market may be determined using the production scheduling model that is used by the Nordic hydropower industry today [5]. In this work, multi–market trade is modelled by including several sale variables in the model, and by letting let each market be represented by a price or premium and an upper limit to the volume that can be traded at the given price. To generate scenario trees for this paper, we use the forecast–based scenario generation method described in [3], and use a set of time–series models to generate the forecasts required as input. The optimization model, however, is general and may be used with any type of scenario–generation method that creates scenarios for the stochastic parameters, i.e. prices and the volume limit.

2 Modelling the Markets

The forecast–based scenario generation method presented in [3] generates scenario trees based on point–forecasts combined with historical forecast errors. We therefore develop a set of time–series models that generates a daily point forecast for the most important properties of the day–ahead and the balancing market. Each market is characterized by a price and a maximum quantity that may be traded at this price. The full presentation of how the markets are modelled by time–series is given in [6], but the most important aspects are repeated here for clarity.

When it comes to the characteristics of each market, the day–ahead market is a daily, centrally cleared auction. Due to the daily clearing of the day–ahead market, the hourly day–ahead market prices cannot really be represented as a pure time–series process. In normal time series, the information set is assumed to be updated when moving from one time step to the next. This is not the case for day–ahead prices because the information set is updated on a daily rather than an hourly basis. Thus, it is more correct to model the day–ahead prices, p_t^D, as a time series of 24–h panel data rather than a single time series. Our method is based on [7], but in addition we account for seasonal variations. Treating the hourly day–ahead prices as panel data allows for modelling correlations between consecutive hours as well as correlations to the same hour on consecutive days.

In regards to the volume limit in the day–ahead market, we assume that the day–ahead market has perfect competition, and that the producer may sell all its output to the market at the given price. The limit on the maximum volume that can be traded is therefore set to be so large that it is never binding for the producer's problem. No separate time–series model is therefore developed for the volume limit in the day-ahead market, it is simply a very large constant for all hours, V^D. The optimization model can take either a deterministic parameter or a stochastic series as input for the volume limit, depending on assumptions on perfect competition or limited liquidity.

Turning to the balancing market, we observe that this market is event–driven, i.e. there is only a demand in the balancing market if there is an imbalance between supply and demand of power. An event is here taken to mean any random event that could not be accurately predicted before closing of the day–ahead market, from power plant failures to line outages to smaller events such as forecasting errors or even structural imbalances. In terms of modelling, the balancing market is described by three properties, namely (i) the balancing state, (ii) the balancing volume and (iii) the balancing price or premium. The balancing state is determined by the real–time balance of supply and demand. If demand exceeds supply the system will need up regulation and vice versa. In fact, the balancing market may be seen as two markets: one for up regulation where the producer offers to ramp up production, and one for down regulation where the producer offers to ramp down. The price for up regulation will be higher than the day–ahead market price, while the price for down regulation will be lower. In an optimization model for multi–market trade, it is the difference between the market prices that are important for the trading strategy. We therefore consider balancing market premiums, ρ_t^{B+} and ρ_t^{B-}, rather than prices. Another benefit of modelling the premiums rather than prices, is that, as found in [6], the premiums may be modelled as independent from the day-ahead prices.

There is demand in the balancing market only if there is an imbalance between supply and demand, and the size of demand is given by the amount of power needed to bring the system back in balance. Due to this limited demand, we model the balancing market by stocahstic trade limits as well as premiums. That is, for each time step, the maximum volume that may be traded is limited by upper bounds, v_t^{B+} and v_t^{B-}. These upper bounds are stochastic parameters in the multi–market optimization problem, and are zero in hours where there is no imbalance.

We thus need a total of four time series to describe the balancing market: premiums and volumes in each direction. However, all of the models for the balancing market are based on one of the models found to have good performance in [8], namely the model based on [9]. This model considers the event-driven nature of the balancing market by using unevenly spaced time–series. In our model, the timing between balancing events is modelled by a moving–average process that is updated every time an event occurs. The balancing volume (i.e. the size of the balancing event if it occurs) is modelled as an autoregressive stationary unevenly spaced time series of order 1. The same type of series is fitted to the balancing market premiums.

The historical data used to fit the time–series for the balancing market describes the total volume of activated power in the balancing market. We must make assumptions on how much of the total volume that can be supplied by an individual agent or hydropower system. One approach is to assume that the individual producer may take a percentage of the total market volume, e.g., 10%, in every hour where there is demand. Another approach is to assume that in each hour, the total market volume may be activated from a single producer with a given probability, e.g., one in ten times. The probability may be related to the number of agents in the power market. In the case studies in Sect. 4, these two approaches are denoted "Percentage" and "Probability".

The process of using the time–series models and the forecast–based scenario–generation method to create input to the optimization model is illustrated in Fig. 1. To develop the time–series models, we use data from Nord Pool's ftp server for the years 2014–2016. For each day of 2017, we then generate a daily forecast with a 72-h forecast horizon. We then use realized data from 2017 to determine historical forecast errors by comparing our daily forecast to historical realized values. This gives us a "database" of historical forecast errors made over a year for forecasting lengths up to 72 h. We only need to initialize this database once before the start of the test period which are the first 18 weeks of 2018. For test instance, the historical forecast errors are used together with new daily point forecasts to generate scenarios. We generate a set of scenarios for the day–ahead market prices and another set of scenarios for the balancing market premiums and volumes. This is because the balancing market premiums and volume are independent of the day–ahead price as explained above. The two set of scenarios are then combined all against all to generate a total scenario tree that describes all the possible outcomes for the day–ahead and balancing market. This total scenario tree is used as input to the stochastic optimization model.

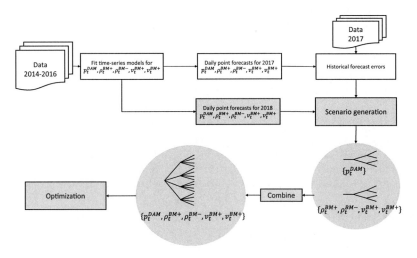

Fig. 1. Process for generating input to the optimization model. The grey boxes are repeated for every test instance, while the white boxes are performed only once to initialize the process.

3   Problem Formulation

This section presents the basic mathematical modellig of the stochastic optimization model that coordinates multi–market trades for a power producer. The producer must determine the trade volumes that maximizes the value of trades made in the day–ahead and the balancing market. This may be expressed as

$$max \sum_t \sum_s \pi_s \left(p_{ts}^D x_{ts}^D + (p_{ts}^D + \rho_{ts}^{B+}) x_{ts}^{B+} + (p_{ts}^D + \rho_{ts}^{B-}) x_{ts}^{B-} \right) \tag{1}$$

where x_{ts}^D, x_{ts}^{B+}, x_{ts}^{B-} are the volumes sold in the day–ahead and the up balancing and down balancing market for delivery at time t in scenario s, and π_s is the probability of scenario s. The need for coordination across markets arises because the final commitment, i.e. the actual volume to be produced in a specific hour, y_{ts}, is the summation over the position made in each market,

$$y_{ts} = x_{ts}^D + x_{ts}^{B+} + x_{ts}^{B-}. \tag{2}$$

In addition, the volumes sold in any market must be less than the demand in the market, i.e.,

$$x_{ts}^D \le V^D, \quad x_{ts}^{B+} \le v_{ts}^{B+} \quad \text{and} \quad x_{ts}^{B-} \le v_{ts}^{B-}. \tag{3}$$

The above model assumes that any volume y_{ts} may be produced. Actual production systems are much more complex. Details of hydropower production are however omitted from the presentation here. In fact, the above model may be used by any producer that participates in the day–ahead and balancing market as long as it is combined with a representation of the specific production system. In our case, the multi–market model is implemented within the framework of models that is used for short–term production scheduling by most large hydropower producers in the Nordic region [5]. The volume to be produced, y_{ts}, is thus determined by this more complex model that includes all technical, hydrological and environmental constraints relevant for hydropower production, e.g., minimum production levels, forbidden operating zones, start/stop, discharge dependent losses in tunnels and penstocks, minimum and maximum reservoir levels, minimum and maximum river or tunnel flows and more.

The simple model formulation above is however not complete without modelling the information structure in the multi–market setting. The day–ahead prices are revealed once every day when the market clears. This means that the scenario tree for day–ahead prices must have a new stage every 24 h. For a 72–h horizon where the scenario tree branches into two new scenarios at each branching step, daily branching would yield $2^2 = 4$ scenarios, see the left part of Fig. 2 for an illustration. In the balancing market, however, prices and volumes are revealed in real–time, which would lead to a scenario tree with hourly branching. This would quickly lead to a very large problem, especially considering that more than two new scenarios at each branching point is necessary to represent the full uncertainty of prices. To avoid this curse of dimensionality, we choose to have daily branching also for the balancing market, i.e., that both the day–ahead prices and the balancing market premiums and volumes are revealed together when the day–ahead market clears. This assumption means that the models sees no uncertainty in the balancing market during each day. This will likely cause an overestimation of the profits obtained by participating in the balancing market because the producer can determine its sales in the balancing market based on known prices and volumes within each day.

In the stochastic program, we use a scenario representation rather than a node formulation, see the right part of Fig. 2. This means that we must explicitly include non–anticipativity constraints stating that if two scenarios s and s' are indistinguishable at time t on the basis of information available at time t, then the decisions made in scenario s must be equal to the decisions made in scenario s'. In our case, this means that the produced volume must be equal between all scenarios belonging to the same node,

$$y_{ts} = y_{ts'}. \tag{4}$$

The same is true for the traded volumes x_{ts}^D, x_{ts}^{B+}, x_{ts}^{B-}. The non–anticipativity constraints are illustrated by the fully drawn grey boxes in Fig. 2. In addition to non–anticipativity related to the daily clearing of the markets, the optimization model also needs to know that trades must be done prior to market clearing and that day–ahead trades must be made prior to balancing market trades. This means that the day–ahead trades cannot depend on any particular realization of prices for the next day. We call this the market non–anticipativity constraints and formulate them as

$$x_{ts}^D = x_{ts'}^D. \tag{5}$$

The market non–anticipativity constraints are illustrated by the dotted grey boxes in Fig. 2. Similar constraints may also be applied to the balancing market trades, x_{ts}^{B+} and x_{ts}^{B-}, depending on whether the trades in the balancing market are to be decided in real–time or not. If the constraint is imposed on the balancing trades, it means that the trades must be determined prior to clearing of the day–ahead market. If the constraints are not imposed, the balancing trades may be determined after clearing of the day–ahead market, i.e. when the producer has knowledge of the realized balancing market prices and volumes. This is of course not possible in reality, but we include it in our case study to measure the value of having perfect information of the balancing market.

Some cases in Sect. 4 also consider the case when the producer is allowed to submit a price–dependent bid curve to the day–ahead market instead of just a single quantity. How the model for short–term hydropower scheduling is extended to also include decisions for optimal bids to the day–ahead market is explained in [4]. In the current framework, determining optimal bids translates to inequality constraints on the volumes traded in the day–ahead market,

$$x_{ts}^D \leq x_{ts'}^D, \quad \text{if} \quad p_{ts}^D \leq p_{ts'}^D, \tag{6}$$

instead of the market non–anticipativity constraints in Eq. (5). This means that the equality constrainst in the dotted grey boxes in Fig. 2 are relaxed to inequality constraints.

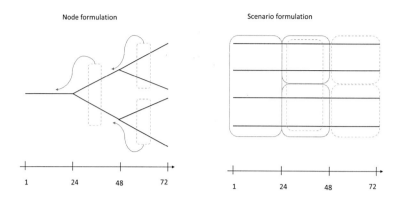

Fig. 2. (Left) Node representation of a scenario tree with daily branching. The dotted grey boxes illustrate that the traded volumes must be determined before prices are revealed. (Right) Scenario representation of a scenario tree with daily branching. The fully drawn grey boxes illustrate the normal non–anticipativity constraints, while the dotted grey boxes represent the market non–anticipativity constraints.

4   Results

In this section, we illustrate how the multi–market model is applied to a simple hydropower system. The system has one reservoir connected to a plant with two generators. The total capacity is 90 MW. We test the optimization model for 30 instances corresponding to initial conditions of 30 different days in the first 18 weeks of 2018. The results given in Tables 1 and 2 are average numbers over the 30 instances. For each instance we use a 72–h horizon with branching in the scenario tree after hour 24 and 48. We use 5 new scenarios at each branching point for the day–ahead prices and 3 scenarios for the balancing market. This results in $5 * 5 * 3 * 3 = 225$ scenarios in total for each instance.

We first consider the case when the producer participates in the day–ahead market only. The next case is when the producer participates in both the day–ahead and balancing market, but consider the two markets sequentially and determines the volumes in the day–ahead market without seeing the balancing market. This is called sequential bidding [1]. In the first case of sequential bidding, we assume that the producer determines all trades in the balancing market at once right after clearing of the day–ahead market. In the second case, we assume that trades in the balancing market may be done in real–time. In the next set of cases, the producer can coordinate its trades in the two markets, that is, the producer may determine the day–ahead bids while also seeing the balancing market. For the coordinated case we also consider cases when balancing market decisions are done only once or in real–time. The different cases are summarized in Table 1, showing the percentage increase in objective function value compared to the base case of participating in the day–ahead market only. We also use different assumptions on the volumes available in the balancing market. Columns 2 and 3 of Table 1 show results when the volume available to the individual producer is 10% and 5% of the total market volume. This means that a low volume is available in most hours. In Columns 4 and 5, however, the total market volume is available to the producer 1 in 10 and 1 in 20 times. This means that large volumes are available in just a few hours. We see that for the 10% and 5% cases, there is a gain in profits from participating in the balancing market. The gain is larger if trades may be coordinated, and even larger if balancing market trades may be done in real–time. The gain of real–time trading is higher than the gain of coordination. The option of trading in real–time corresponds to having perfect information about the balancing market, which is not possible in reality. For the 1/10 and 1/20 cases, there is a gain of real–time trading but not from coordinating trades. This is because balancing markets volumes are so rare that they do not influence the trading strategy if they are to be determined prior to operations.

Table 1. Percentage increase in objective function value compared to participating the day–ahead market only. The producer submits only one production volume to the market.

Case	Percentage		Probability	
	10%	5%	1/10	1/20
DAM	-	-	-	-
DAM + BM sequential	1.12	0.56	-	-
DAM + BM sequential + real–time	1.56	0.82	0.20	0.10
DAM + BM coordinated	1.32	0.55	-	-
DAM + BM coordinated + real–time	2.73	1.28	0.20	0.10

We repeat the same cases as above, but now we assume that the producer can submit a price–dependent bid curve to the day–ahead market. The results are summarized in Table 2. In general, we see similar results as in the case without bidding: there is a gain from coordinating trades and an even larger gain if balancing markets trades may be done in real–time, i.e. with perfect information. Another result, although not evident from the tables, is that the objective function value when submitting bids is higher than when the producer submits just a single quantity, i.e., the base case of participating in only the day–ahead market is 0.88% higher in Table 2 than in Table 1. This is because the market non–anticipativity constraints (equality constraints between scenarios) are relaxed to inequality constraints in the bidding problem.

Table 2. Percentage increase in objective function value compared to participating the day–ahead market only. The producer submits a price-dependent bid curve to the market.

Case	Percentage		Probability	
	10%	5%	1/10	1/20
DAM	-	-	-	-
DAM + BM sequential	0.21	0.10	-	-
DAM + BM sequential + real–time	2.07	1.14	0.19	0.06
DAM + BM coordinated	1.19	0.46	-	-
DAM + BM coordinated + real–time	2.66	1.30	0.19	0.07

5   Conclusions

This paper has presented a stochastic program that illustrates how the information structure in a multi–market setting may be modelled for a power producer participating in the day–ahead and balancing power market. The formulation includes normal non–anticipativity constraints that represents the daily clearing of the markets. We also include market non–anticipativity constraints which represents that trades must be made prior to market clearing and that day–ahead trades must be made prior to balancing market trades. Similar restrictions may be applied to the balancing market trades, depending on whether they are to be determined in real–time or not. We find that there is a gain from coordinating trades across markets, and an even larger gain from bidding in the balancing market in real–time. The option of trading in real–time corresponds to having perfect information about the balancing market, which is not possible in reality. We also use two different methods for representing the balancing market volume that is available to an individual producer: either the producers see a small percentage of the total market volume in all hours, or the entire market volume is available for the producer with a given probability. The probability may be based on the number of agents in the market and may thus be a realistic representation of the balancing market volume.

References

1. Boomsma, T.K., Juul, N., Fleten, S.-E.: Bidding in sequential markets: the Nordic case. Eur. J. Oper. Res. **238**(3), 797–809 (2014). https://doi.org/10.1016/j.ejor.2014.04.027
2. Birge, J.R., Louveaux, F.: Introduction to Stochastic Programming. Springer, New York (2011)
3. Kaut, M.: Forecast–based scenario–tree generation method. Optimization Online (2017)
4. Aasgård, E.K., Naversen, C.Ø., Fodstad, M., Skjelbred, H.I.: Optimizing day-ahead bid curves in hydropower production. Energy Syst. (2017). https://doi.org/10.1007/s12667-017-0234-z

5. SINTEF Energy Research SHOP. https://www.sintef.no/en/software/shop/. Accessed 4 Apr 2018
6. Aasgård, E.K.: Modelling prices in sequential electricity markets, working paper at NTNU (2018)
7. Huismann, R., Huurmann, C., Mahieu, R.: Hourly electricity prices in day-ahead markets. Energy Econ. **29**, 916–928 (2007). https://doi.org/10.1016/j.eneco.2006.08.005
8. Klæboe, G., Eriksrud, A.L., Fleten, S.-E.: Benchmarking time series based forecasting models for electricity balancing market prices. Energy Syst. **6**(1), 43–61 (2015). https://doi.org/10.1007/s12667-013-0103-3
9. Croston, J.D.: Forecasting and stock control for intermittent demands. Oper. Res. Q. **23**, 289–303 (1972). https://doi.org/10.2307/3007885

Assessing the Impacts of Integrating Snowpack Error Distribution in the Management of a Hydropower Reservoir Using Bayesian Stochastic Dynamic Programming (BSDP)

Richard Arsenault[1(✉)] [iD], Pascal Côté[2], and Marco Latraverse[2]

[1] Department of Construction Engineering, École de Technologie Supérieure, Montréal, Canada
Richard.arsenault@etsmtl.ca
[2] Rio Tinto Aluminum - Power Operations, Jonquière, Canada

Abstract. A hydropower system is presented in which long-term hydrological forecasts must be performed. The system is strongly snowmelt-dominated and the duration of the spring flood can last upwards of 5 months. The risk of flooding is very high when the snowpack is above long-term average values. This work analyzes the impacts of estimating and integrating the snowpack error distribution in a hydrological forecasting framework when optimized by a Bayesian Stochastic Dynamic Programming (BSDP) reservoir management optimization algorithm. The methodology follows two main steps. In the first step, the hydrological model is run on the historical dataset. The resulting hydrograph and hydrologic states are then compared to those of a synthetic "perfect" model simulation. An error distribution is defined between both series that can be used in the BSDP framework. Second, the first step is repeated with a classical SDP approach instead of BSDP to quantify the impacts of using the error distribution on hydropower generation. Results show that BSDP outperforms the classical SDP and that the snowpack error estimation plays a significant role in improving the reservoir management policy.

Keywords: Bayesian Stochastic Dynamic Programming
Snow water equivalent · Hydrological forecasting

1 Introduction

Hydropower reservoir management is a complex task given the uncertain natural processes that drive the decision process and the fixed constraints governing the reservoir operations. Decision support systems have been developed to guide reservoir managers and aid them in making the best possible decisions based on the current states and expected future inflow realizations. However, the stochastic nature of the underlying processes makes it a difficult proposition. Stochastic optimization methods, such as the Stochastic Dynamic Programming (SDP) approach, were developed to consider this uncertainty [1]. Variants of SDP were proposed, such as the Sampling SDP (SSDP), in which inflow scenarios are used instead of purely statistical distributions of

© Springer Nature Switzerland AG 2019
A. Helseth (Ed.): HSCM 2018, *Proceedings of the 6th International Workshop on Hydro Scheduling in Competitive Electricity Markets*, pp. 26–32, 2019.
https://doi.org/10.1007/978-3-030-03311-8_4

inflows [2], and Stochastic Dual Dynamic Programming (SDDP) in which states are sampled based on a forward simulation step and sub-problems are linearized to estimate the multivariate water value function using Benders cuts [3]. In this paper, we discuss another variant called Bayesian SDP (BSDP) in which the uncertainty in the inflow forecasts are informed by an evolving error function whose posterior distribution is updated in time [4]. It has been widely applied in reservoir and multi-reservoir management [5–7]. In this study, the Bayesian component is evaluated on the snowpack as an indicator of future inflows rather than the usual error around the forecasted inflows.

2 Study Site

The study was performed on the Nechako River basin upstream of the Skins Lake Spillway, in British-Columbia, Canada, as shown in Fig. 1. The catchment is owned and operated by Rio Tinto's Aluminum division. Rio Tinto is a global leader in mining and transformation of primary metals such as iron, aluminum and copper. The Aluminum branch owns and operates electric generating stations to provide the necessary energy for the aluminum smelting process.

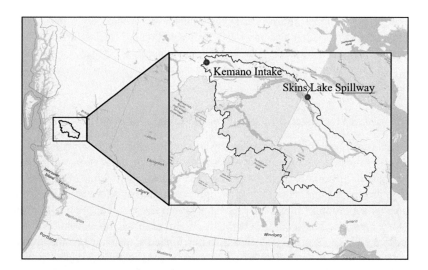

Fig. 1. Nechako Reservoir owned and operated by Rio Tinto in British-Columbia, Canada, for hydropower generation. The two reservoir outlets are shown here, namely the Kemano Intake for the hydropower generating station on the West side and the Skins Lake Spillway on the East side.

One of these stations is the Kemano hydropower generating station on the study site, which powers the Kitimat smelter. The Nechako reservoir has two outlets: one at the Kemano power plant, which drains into the Pacific Ocean, and the other at the Skins Lake Spillway on the east side, which feeds the Nechako River that is lined with villages and is prone to flooding. The main risk in the operation of the Nechako

Reservoir is causing flooding to these villages. Therefore, long-term ensemble forecasts are produced to generate probable inflow scenarios and an optimization algorithm is used to determine the optimal actions. The catchment has a total area of 14040 km^2 and sees average inflows of approximately 200 m^3/s. Operational constraints include reserved environmental flows from the Skins Lake Spillway, maximizing generation at the Kemano powerhouse and minimizing floods in downstream villages. The environmental flows are also modulated according to the period of the year to accommodate different species such as spawning salmons and white sturgeons.

3   Data and Methods

3.1   Data

The study makes use of Rio Tinto's hydrometeorological observation records, including daily air temperature, precipitation and reservoir inflows from 1957–2017. Precipitation is highly dependent on topography and can range from less than 500 mm/year in the drier areas to over 2000 mm in the mountain ranges.

3.2   Hydrological Model

To take these variations into account, these data were used to drive the CEQUEAU hydrological model, which was previously calibrated on the Nechako river basin using the measured reservoir inflows as the calibration target. CEQUEAU is a distributed model that can estimate the Snow Water Equivalent (SWE) and soil moisture variables across the study site [10].

3.3   Bayesian Stochastic Dynamic Programming Algorithm

The Stochastic Dynamic Programming (SDP) optimization algorithm is a feedback-type explicit method which requires solving a series of optimization sub-problems to find an operating policy which is represented by a series of functions $x_t = G_t(y(t))$ where y_t are the system states at the beginning of the period and x_t are the optimal decisions at time step t [4, 8]. The challenge lies in the fact that the decision taken in time step t must be taken according to the uncertainties of the inflows and will influence the optimal decision in time step $t+1$ due to the impacts x_t will have on the system state y_{t+1}. In this study, the system states y_t are composed of two variables, namely the reservoir storage (s) and a hydrological variable (h). The optimal policy will then depend on both the current reservoir storage and the hydrological variable which is a predictor of future inflows [5]. The Nechako Reservoir is strongly dominated by snow, and as such, snowmelt is both highly influential and a major driver of the water management policy. Therefore, snowpack depth combined with soil humidity averaged over the basin is a generally good predictor of future inflows and is used as the hydrological variable in this study [9].

In this context the SDP algorithm consists in finding a water value function $F_t()$ for each time step, as follows:

$$F_t(s_t, h_t) = \max_{x_t} \left\{ \mathop{E}_{q_t|h_t} \left[B_t(x_t, s_t, q_t) + \mathop{E}_{h_{t+1}|h_t} [F_{t+1}(s_{t+1}, h_{t+1})] \right] \right\} \tag{1}$$

and where q_t is the natural inflow to the reservoir.

The Bayesian component to the BSDP algorithm comes from the addition of an estimator of error around the hydrological variable. Instead of having a single deterministic value for the hydrological variable, a distribution of likely values is used to inform the optimal decision. In an operational context, the measured hydrological variable $\hat{h}_t$ will be an estimator of h_t where $h_t = \hat{h}_t + \varepsilon_t$ and where $\varepsilon_t \sim d_{t,\theta}$ and $d_{t,\theta}$ is the distribution of the error that must be updated at the end of each season. The idea behind BSDP is to solve Eq. (1) in the SDP optimization step using the historical data and to simulate the following policy to evaluate the performance of the system or to find the best decision to apply in real time given the current storage s_t and the measured hydrological variable $\hat{h}_t$:

$$x_t^* = \arg\max_{x_t} \left\{ \mathop{E}_{h_t|\hat{h}_t} \left[\mathop{E}_{q_t|h_t} \left[B_t(x_t, s_t, q_t) + \mathop{E}_{h_{t+1}|h_t} [F_{t+1}(s_{t+1}, h_{t+1})] \right] \right] \right\} \tag{2}$$

3.4   Methodology

The methodology employed to assess the usefulness of snowpack error in optimizing the hydropower generation followed three main steps. First, two simulation scenarios were generated, as will be detailed later. Second, the scenarios were compared to establish error distributions between them. Finally, the scenarios along with the error distributions were used in the BSDP algorithm and the impacts of the optimal decisions were tested in a hydropower simulator. This allowed quantifying the overall energy generation and flood risk of using the BSDP method. For comparison, a classical SDP was also implemented and did not make use of the snowpack error estimations.

The two simulated inflow scenarios introduced previously will now be detailed. In both cases, the inflows as well as a custom hydrologic variable (average available water in the model's soil, underground and snowpack reservoirs), which is used by BSDP to estimate future inflow volumes, are generated. In the first scenario, the hydrologic model was run with no particular preprocessing, generating a base-case scenario of simulated inflows to the reservoir for the entire study period (1957–2017). The second simulation was performed by driving the CEQUEAU model with a specially constructed set of modified precipitation and temperature values. This set of precipitation and temperature values, when run in the CEQUEAU model, generates inflow scenarios that perfectly match the observed inflows. In essence, they form a pseudo-perfect set of climate variables. This allowed running CEQUEAU to extract the hydrological variable associated with a perfect inflow simulation, i.e. the "real-world" hydrologic variable.

The error distribution was computed between the pseudo-perfect simulation hydrologic variable and that of the base-case simulation for each day in the winter period. These distributions were used as the a priori estimates in the BSDP algorithm. The BSDP was then used to estimate the best decision to take for maximizing revenue from aluminum smelting and energy exports while respecting multiple constraints such as dam safety, flooding risk and environmental release requirements. The methodology was also performed without feeding the error distribution to the BSDP algorithm to evaluate the value of the added information content, which amounts to performing a classical SDP approach.

4   Results and Analysis

The first interesting results are the error distributions for the different periods of the year. In this study, the error distributions are defined for each day of the year, i.e. there are 365 error distributions. Figure 2 shows the aggregate of these error distributions for the entire year (top panel), the winter days (bottom left) and remainder and summer and fall days (bottom right). It is clear that the majority of the error in inflow volumes stems from the winter days, which is consistent with the difficulties in measuring snow depth and snow water equivalent (SWE) in remote catchments.

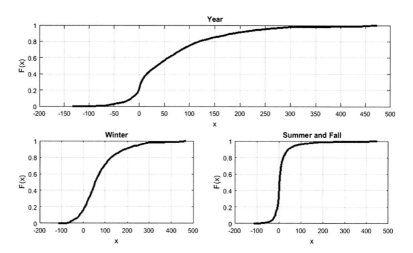

Fig. 2. Distributions of the error in the hydrological variable between the pseudo-perfect simulation and the original, base-case simulation. Units are in millimeters of water.

Then, the BSDP and SDP algorithms were fed with the simulated inflows and the BSDP was also given the error distribution.

The reservoir level in Fig. 3 is maintained slightly lower with BSDP. This might be because the error distribution takes into account a bias in the measured snow precipitations, possibly due to systematic precipitation underestimation. Therefore, the policy might favor generating more water, knowing that there are higher-than-expected inflows incoming. Finally, the hydropower simulator allowed showing that for identical shortage risk, the BSDP algorithm allowed generating 3% more hydropower than the SDP algorithm while reducing the probability of flooding by 2.5%. Unfortunately, the actual GWh values are proprietary and cannot be divulged in absolute terms.

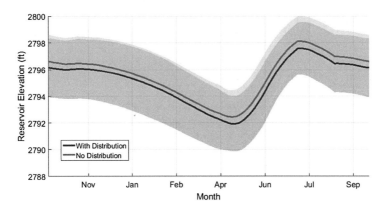

Fig. 3. Reservoir level with (BSDP) and without (SDP) the hydrologic variable error distributions. The translucent envelopes around each curve represent the standard deviation around the value over the 60-year period.

5 Conclusion

Rio Tinto, as well as other hydropower utilities and industries, must constantly improve their methods to ensure dam and population safety while maximizing the efficiency of their installations. This study aims to combine the proven BSDP reservoir optimization algorithm with a novel hydrologic variable error determination technique to generate improved reservoir operating decisions. We show that the integration of error distributions possesses value and expect that continuously updating the posterior error distribution will further improve the decision-making process. It is expected that implementing this framework will allow Rio Tinto to increase its efficiency and reduce the risk of floods on the Nechako River.

References

1. Stedinger, J.R., Sule, B.F., Loucks, D.P.: Stochastic dynamic programming models for reservoir operation optimization. Water Resour. Res. **20**(11), 1499–1505 (1984)
2. Faber, B.A., Stedinger, J.R.: Reservoir optimization using sampling SDP with ensemble streamflow prediction (ESP) forecasts. J. Hydrol. **249**(1–4), 113–133 (2001)
3. Shapiro, A.: Analysis of stochastic dual dynamic programming method. Eur. J. Oper. Res. **209**(1), 63–72 (2011)
4. Karamouz, M., Vasiliadis, H.V.: Bayesian stochastic optimization of reservoir operation using uncertain forecasts. Water Resour. Res. **28**(5), 1221–1232 (1992)
5. Tejada-Guibert, J.A., Johnson, S.A., Stedinger, J.R.: Comparison of two approaches for implementing multi-reservoir operating policies derived using stochastic dynamic programming. Water Resour. Res. **29**(12), 3969–3980 (1993)
6. Kim, Y.O., Palmer, R.N.: Value of seasonal flow forecasts in Bayesian stochastic programming. J. Water Resour. Plan. Manag. **123**(6), 327–335 (1997)
7. Mujumdar, P.P., Nirmala, B.: A Bayesian stochastic optimization model for a multi-reservoir hydropower system. Water Resour. Manag. **21**(9), 1465–1485 (2007)

8. Côté, P., Leconte, R.: Comparison of stochastic optimization algorithms for hydropower reservoir operation with ensemble streamflow prediction. J. Water Resour. Plan. Manag. **142** (2), 04015046 (2015)

9. Desreumaux, Q., Côté, P., Leconte, R.: Role of hydrologic information in stochastic dynamic programming: a case study of the Kemano hydropower system in British Columbia. Can. J. Civ. Eng. **41**(9), 839–844 (2014)

10. Arsenault, R., Latraverse, M., Duchesne, T.: An efficient method to correct under-dispersion in ensemble streamflow prediction of inflow volumes for reservoir optimization. Water Resour. Manag. **30**(12), 4363–4380 (2016)

Inflow Forecasting for Hydropower Operations: Bayesian Model Averaging for Postprocessing Hydrological Ensembles

Andreas Kleiven[1,2(✉)] and Ingelin Steinsland[1]

[1] Department of Mathematical Sciences,
Norwegian University of Science and Technology, NO-7491, Trondheim, Norway
andreas.kleiven@ntnu.no
[2] Department of Industrial Economics and Technology Management,
Norwegian University of Science and Technology, NO-7491, Trondheim, Norway

Abstract. This paper contributes to forecasting of renewable infeed for use in dispatch scheduling and power systems analysis. Ensemble predictions are commonly used to assess the uncertainty of a future weather event, but they often are biased and have too small variance. Reliable forecasts for future inflow are important for hydropower operation, and the main purpose of this work is to develop methods to generate better calibrated and sharper probabilistic forecasts for inflow. We propose to extend Bayesian model averaging with a varying coefficient regression model to better respect changing weather patterns. We report on results from a case study from a catchment upstream of a Norwegian power plant during the period from 24 June 2014 to 22 June 2015.

Keywords: Bayesian model averaging
Probabilistic postprocessing · Inflow forecasting

1 Introduction

Hydrological forecasting plays an important role in a variety of applications, ranging from flood prevention to water resource management and hydropower production. Forecasting inflow to hydropower reservoirs for operation and scheduling is the focus of this work. Future streamflows are uncertain, and forecasts generated from hydrological models are subject to errors. In order to quantify the uncertainty of future streamflows, it is common to generate an ensemble of forecasts with perturbations made for both the initial state and the model formulation for each member of the ensemble. The resulting ensemble can be interpreted as a probabilistic forecast. However, the ensemble forecasts tend to be underdispersive, meaning that the observed value too often lies outside the

© Springer Nature Switzerland AG 2019
A. Helseth (Ed.): HSCM 2018, *Proceedings of the 6th International Workshop on Hydro Scheduling in Competitive Electricity Markets*, pp. 33–40, 2019.
https://doi.org/10.1007/978-3-030-03311-8_5

ensemble range. Therefore, statistical postprocessing methods are essential in order to obtain calibrated and sharp probabilistic forecasts.

A widely used postprocessing methodology for ensemble forecasts is Bayesian model averaging (BMA) [1]. In the BMA methodology, a component probability density function (pdf) is assigned to each ensemble member forecast, and the BMA probabilistic forecast is given by a weighted average of the individual ensemble member pdfs. Another popular postprocessing method is Ensemble model output statistics (EMOS) [2]. This method is based on multiple linear regression. An advantage with the BMA methodology is that the method respects the dynamics in the ensemble.

In the original BMA approach for postprocessing of forecast ensembles, a Gaussian pdf is assigned to the ensemble members [1]. Extensions of the BMA methodology have been developed for cases where the dependent variable deviates from the Gaussian case. Sloughter et al. [3] modified the method to apply to precipitation forecasts by introducing a discrete-continuous model which combines a logistic regression model and gamma distributions. Moreover, BMA using gamma distributions as the component pdfs has been applied to wind speed forecasting [4]. Furthermore, Duan et al. [5] used the BMA approach to generate probabilistic hydrological forecasts after transforming streamflow values using the Box-Cox transformation.

In this paper we aim to generate reliable probabilistic forecasts for inflow by extending the original BMA methodology. Many stochastic optimization methods used for operational purposes often require a large number of inflow scenarios as input, and inflow forecasts in the form of predictive distributions are useful in the sense that one easily can generate many inflow scenarios from sampling. Séguin et al. [6] propose a method for the natural next step of our analysis, which is a transition from a probabilistic forecast to a scenario tree or a lattice, something that is useful for input in short-term hydropower operational optimization methods. We propose to extend the BMA methodology for ensemble forecasts with varying coefficient regression (VCR) [7]. We demonstrate the method in a case study from a catchment upstream of a Norwegian power plant during the period from 24 June 2014 to 22 June 2015.

2   Bayesian Model Averaging Using Varying Coefficient Regression

The use of BMA for statistical postprocessing of forecast ensembles was introduced by Raftery et al. [1]. The BMA approach generates a probabilistic forecast in the form of a predictive pdf by combining deterministic forecasts from different models. We suggest to extend the BMA methodology by using a VCR model, which we denote BMA-VCR. The models presented below can be applied to each lead time individually, where lead time refers to the forecast horizon.

We assume that the ensemble members are exchangeable, meaning that they are treated equally. Therefore, we present the BMA methodology for exchangeable member forecast. First, we follow the approach of Raftery et al. [1] and consider the normal distribution with mean $\alpha + \beta x_m$ and standard deviation τ as the ensemble member pdfs. The BMA probabilistic forecast is then given by

$$f(y|x_1, ..., x_M) = \frac{1}{M} \sum_{m=1}^{M} g(y|x_m)$$

$$Y|x_m \sim \mathcal{N}(\mu_m, \tau^2)$$

$$\mu_m = \alpha + \beta x_m, \tag{1}$$

where x_m is the deterministic forecast from ensemble member m, Y is the random variable representing future inflow to be forecasted, and M is the size of the ensemble. The bias-correction parameters, α and β, are equal for each ensemble member and the weights for exchangeable member forecasts are $\frac{1}{M}$. However, such a simple linear bias-correction does in general not provide good predictions for heteroskedastic and non-Gaussian model errors, which is likely to occur in hydrological forecasting [8]. To easier incorporate local weather patterns, we suggest to apply a nonlinear bias-correction in the form of a VCR model. VCR models are a class of generalized linear regression models where the coefficients are allowed to vary as functions of other variables. We let the BMA bias-correction parameters vary throughout time t, and a VCR model can then be described by

$$\alpha_t = \alpha_{t-1} + a_t, \quad a_t \sim N(0, \delta^{-1})$$

$$\beta_t = \beta_{t-1} + b_t, \quad b_t \sim N(0, \delta^{-1}), \tag{2}$$

where we restrict the precision parameter δ to be equal for both processes. We refer to δ as a precision parameter since the larger value, the less variance. In the BMA-VCR model, we include both static bias-correction parameters α and β and dynamic parameters α_t and β_t, which leads to the following form of the BMA-VCR probabilistic forecast

$$f(y|x_1, ..., x_M) = \frac{1}{M} \sum_{m=1}^{M} g(y|x_m)$$

$$Y|x_m \sim \mathcal{N}(\mu_m, \tau^2)$$

$$\mu_m = (\alpha + \alpha_t) + (\beta + \beta_t)x_m$$

$$\alpha_t = \alpha_{t-1} + a_t, \quad a_t \sim N(0, \delta^{-1})$$

$$\beta_t = \beta_{t-1} + b_t, \quad b_t \sim N(0, \delta^{-1}). \tag{3}$$

The static parameters α and β represent the total bias between forecast and observation pairs from a training period, and the dynamic parameters α_t and β_t evolve from time $t = 1$. The parameter δ decides the flexibility of the dynamic parameters. A small δ gives a good fit to training data, but generally not good predictions, i.e. overfitting, while a large value of δ gives less flexibility for the dynamic parameters. By letting $\delta^{-1} = 0$ the BMA-VCR model formulation in (3) coincide with the original BMA model defined in (1). The assumption that δ^{-1} is identical for α_t and β_t is reasonable as the dependency between the estimators for α_t and β_t is high. If $\alpha + \alpha_t = 0$, and $\beta + \beta_t = 1$, the mean of an ensemble member forecast, μ_m, will be the deterministic forecast x_m. If $\beta + \beta_t < 1$, we expect $\alpha + \alpha_t > 0$. Furthermore, if $\beta + \beta_t > 1$, we expect $\alpha + \alpha_t < 0$.

3   Parameter Estimation

In the original BMA methodology for ensemble forecasts, a sliding window of constant size, consisting of forecast and observation data from the most recent history, is used to train the model. In the VCR models, there are dynamic parameters that evolve from time $t = 1$. We use Bayesian inference for estimation of bias-correction parameters $\alpha, \beta, \alpha_t, \beta_t$. The variance parameter τ is then estimated from a sliding window training period in the same way as in the work of Raftery et al. [1]. The last model parameter, the precision parameter δ, is estimated based on predictive performance. For inference, we apply integrated nested Laplace approximations (INLA) [9,10]. INLA is a method for performing approximate Bayesian inference. As an alternative to simulation-based Monte Carlo integration, INLA uses the analytic approximation with the Laplace method, which leads to computational benefits. Furthermore, R-INLA [11], which is an open source software, is suitable for parameter estimation in the BMA-VCR model.

4   Forecast Verification

Probabilistic forecasts take the form of predictive pdfs, and in order for the forecast to be useful, it is important to assess the predictive performance. The models are evaluated according to calibration and sharpness. Calibration is the statistical consistency between the predictive pdfs and the corresponding observed values. Sharpness is a measure of uncertainty of the predictive pdfs.

The verification rank histogram (VRH) is often used to assess calibration of ensemble forecasts [1,3,12]. The VRH is computed by arranging the ensemble forecasts and the corresponding observation in increasing order. To assess calibration of probabilistic forecasts, the probability integral transform (PIT) is common to apply [1,3,13]. The probabilistic forecast is calibrated if the PIT values, which is the value of the predictive cdf at the corresponding observation, are uniformly distributed. Uniformity can be assessed by making a histogram of PIT values. The shape of the VRH and the PIT histogram, gives an indication whether the probabilistic forecast is calibrated. Hump-shaped histograms indicate that the probabilistic forecast is overdispersed, which means that the prediction intervals on average are too wide. U-shaped histograms indicates underdispersion, meaning that the prediction intervals on average are too narrow. Asymmetrical histograms occur when the probabilistic forecast is biased.

Proper scoring rules are often used to assess the predictive performance of a probabilistic forecast. A scoring rule is proper if the expected score is minimized when the issued forecast is the true distribution of the quantity to be forecasted [14]. The continuous ranked probability score (CPRS) is a proper scoring rule that measures both calibration and sharpness of a probabilistic forecast [15]. The CRPS measures the difference between the predicted and occurred cumulative distributions. The value of the CRPS is non-negative and the smaller value the better quality of the probabilistic forecast. For deterministic forecasts, the CRPS reduces to the absolute error, hence it is possible to compare the performance of probabilistic forecasts and deterministic forecasts.

5   Data and Study Area

The ensemble forecasts used in this study are generated from the Hydrologiska Byråns Vattenbalansavdelning (HBV) model [16]. The model has a number of free hydrological parameters that are estimated from training data, and the start state is estimated using observed precipitation and temperature from the history. Ensembles of temperature and precipitation forecasts from the European Centre for Medium-Range Weather Forecasts (ECMWF) are used as input in the HBV-model. The ensemble size in this study is $M = 51$, and the ensemble forecasts are treated equally, i.e. they are exchangeable.

In the case study, we consider the Osali catchment which is a part of the Ulla-Førre hydropower complex south west in Norway [17,18]. Daily inflow observations, in unit $m^3\,s^{-1}$, are recorded and data are provided by Statkraft, which is the largest hydropower producer in Norway. The method is evaluated for lead time $l = 1$ day, where lead time refers to the forecast horizon.

6   Results and Discussion

We apply the BMA-VCR method to inflow forecasting from the Osali catchment. The method is tested in the period from 24 June 2014 to 22 June 2015. We analyze how the precision parameter δ influence the predictive performance by considering mean CRPS, which is the average CRPS taken over all days in the period under study. The lower mean CRPS, the better predictive performance. Calibration is assessed through the PIT histogram.

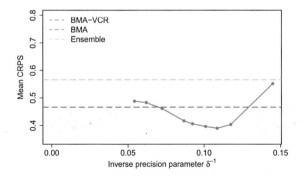

Fig. 1. Mean CRPS as a function of the inverse precision parameter δ^{-1}. The figure shows the potential of including a VCR model in the BMA methodology for postprocessing of hydrological ensembles. We observe that mean CRPS is lower for BMA-VCR compared to BMA with static parameters for certain values of δ^{-1}.

Mean CRPS is given as a function of the inverse precision parameter δ^{-1} in Fig. 1. The original BMA method is the purple horizontal line, and the mean CRPS of the ensemble is given by the green line. The BMA-VCR method, which corresponds to a non-linear bias-correction in the original BMA methodology is shown in red. Where lines intersect means that the predictive performance is equally good. We observe that a large value of δ^{-1} leads to large mean CRPS, which in this case corresponds to overfitting and poor predictive performance. We observe that an inverse precision parameter value close to 0.11 provides a good forecasting performance for the BMA-VCR method. We get mean CRPS values 0.57, 0.47, and 0.39 for the raw ensemble, BMA method, and BMA-VCR method respectively.

We observe from Fig. fig:crpsspsdelta that the right choice for δ^{-1} is important. Values between $\delta^{-1} = 0.07$ and $\delta^{-1} = 0.13$ leads to better predictive performance compared to the original BMA method. Values outside this interval leads higher CRPS values. The computation time for estimating parameters in the BMA-VCR method, using Bayesian inference, is longer compared to the BMA method, which uses maximum likelihood estimation. The computation time increases with increasing amount of data used for fitting the model. With small amount of data available, the computation time of the BMA-VCR method is similar to the original BMA method. As more data become available, the estimation procedure takes longer time. For operational use, a sliding window with constant size can be applied to reduce computation time.

The PIT histogram obtained for the probabilistic forecast from the BMA-VCR method and the VRH for the ensemble forecasts are provided in Fig. 2. The horizontal dotted line indicate the height of the bars for a perfectly calibrated forecast. We observe that the VRH for lead time 1 is strongly u-shaped and slightly biased. This means that the ensemble underestimate variance. The PIT histogram obtained from the probabilistic forecast of the BMA-VCR method is closer to uniform.

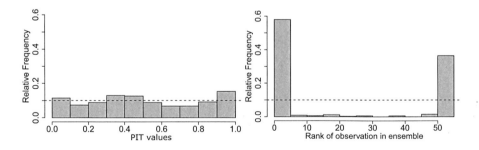

Fig. 2. He PIT histogram obtained from the probabilistic forecast of the BMA-VCR method (left) and the VRH from the raw ensemble (right)

The method can be further extended and applied to a higher-dimensional system, but this is not tested in this work. For multiple catchments, the dependency between corresponding ensemble members will be handled by the current method. However, further extensions to the proposed methodology are needed.

7   Conclusion

In this work we have presented a new postprocessing method for hydrological ensembles. We have suggested to extend the original BMA approach for postprocessing of ensemble forecasts with a VCR model. The performance of the postprocessing methods was demonstrated in a case study of the Osali catchment in the south-western part of Norway for lead time $l = 1$ day. The results showed that applying a non-linear regression for the bias-correction parameters in the original BMA methodology has great potential to improve the predictive performance of hydrological ensembles for short lead times.

Acknowledgements. The authors would like to thank Stein-Erik Fleten for valuable discussions. We also thank Statkraft and Stian Solvang Johansen for providing data and motivation behind this work. Andreas Kleiven acknowledges support from the Research Council of Norway, through HydroCen, project number 257588. For Ingelin Steinsland this project was supported by the Research Council of Norway, project 250362.

References

1. Raftery, A.E., Gneiting, T., Balabdaoui, F., Polakowski, M.: Using Bayesian model averaging to calibrate forecast ensembles. Monthly Weather Rev. **133**(5), 1155–1174 (2005)
2. Gneiting, T., Raftery, A.E., Westveld, A.H., Goldman, T.: Calibrated probabilistic forecasting using ensemble model output statistics and minimum CRPS estimation. Monthly Weather Rev. **133**(5), 1098–1118 (2005)
3. Sloughter, J.M.L., Raftery, A.E., Gneiting, T., Fraley, C.: Probabilistic quantitative precipitation forecasting using Bayesian model averaging. Monthly Weather Rev. **135**(9), 3209–3220 (2007)
4. Sloughter, J.M., Gneiting, T., Raftery, A.E.: Probabilistic wind speed forecasting using ensembles and Bayesian model averaging. J. Am. Stat. Assoc. **105**(489), 25–35 (2010)
5. Duan, Q., Ajami, N.K., Gao, X., Sorooshian, S.: Multi-model ensemble hydrologic prediction using Bayesian model averaging. Adv. Water Resour. **30**(5), 1371–1386 (2007)
6. Séguin, S., Fleten, S.E., Côté, P., Pichler, A., Audet, C.: Stochastic short-term hydropower planning with inflow scenario trees. Eur. J. Oper. Res. **259**(3), 1156–1168 (2017)
7. Hastie, T., Tibshirani, R.: Varying-coefficient models. J. R. Stat. Soc. Ser. B (Methodol.) **6**(3), 757–796 (1993)
8. Vrugt, J.A., Robinson, B.A.: Treatment of uncertainty using ensemble methods: comparison of sequential data assimilation and Bayesian model averaging. Water Resour. Res. **43**(1) (2007)

9. Rue, H., Martino, S., Chopin, N.: Approximate Bayesian inference for latent Gaussian models by using integrated nested laplace approximations. J. R. Stat. Soc.: Ser. B (Stat. Methodol. **71**(2), 319–392 (2009)
10. Blangiardo, M., Cameletti, M.: Spatial and Spatio-Temporal Bayesian Models with R-INLA. John Wiley & Sons, Hoboken (2015)
11. Martins, T.G., Simpson, D., Lindgren, F., Rue, H.: Bayesian computing with INLA: new features. Comput. Stat. Data Anal. **67**, 68–83 (2013)
12. Hamill, T.M.: Interpretation of rank histograms for verifying ensemble forecasts. Monthly Weather Rev. **129**(3), 550–560 (2001)
13. Dawid, A.P.: Present position and potential developments: some personal views: statistical theory: the prequential approach. J. R. Stat. Soc. Ser. A (Gen.) **147**(2), 278–292 (1984)
14. Gneiting, T., Raftery, A.E.: Strictly proper scoring rules, prediction, and estimation. J. Am. Stat. Assoc. **102**(477), 359–378 (2007)
15. Hersbach, H.: Decomposition of the continuous ranked probability score for ensemble prediction systems. Weather Forecast. **15**(5), 559–570 (2000)
16. Bergström, S.: The HBV model: its structure and applications. Swedish Meteorological and Hydrological Institute (1992)
17. Engeland, K., Steinsland, I.: Probabilistic postprocessing models for flow forecasts for a system of catchments and several lead times. Water Resour. Res. **50**(1), 182–197 (2014)
18. Engeland, K., Steinsland, I., Johansen, S.S., Petersen-Øverleir, A., Kolberg, S.: Effects of uncertainties in hydrological modelling. A case study of a mountainous catchment in Southern Norway. J. Hydrol. **536**, 147–160 (2016)

Benchmarking Hydro Operation by Use of a Simulator

Birger Mo[1(✉)], Sara Martino[1], Christian Naversen[1],
Gunnar Aronsen[2], and Ole Rismark[3]

[1] SINTEF Energy Research, Sem Sælands vei 11, 7034 Trondheim, Norway
birger.mo@sintef.no
[2] TrønderEnergi, Klæbuveien 118, 7031 Trondheim, Norway
[3] SKM Market Predictor AS,
Olav Tryggvasons Gate 2B, 7011 Trondheim, Norway

Abstract. The paper describes how a simulator is used to benchmark TrønderEnergis' historical operation of one of their hydro systems. The simulator is a data program that simulates daily hydro optimization and scheduling tasks for the historical period 2005 to 2015. The purpose of the benchmark is to evaluate how good the historical operation has been and to point to which tasks in the decision process that is most important to improve (e.g. price forecasting, inflow forecasting or snow storage information).

Keywords: Price forecasting · Inflow forecasting · Hydro optimization
Water values

1 Introduction

The paper describes results from a project where the goal was to benchmark TrønderEnergis' historical operation of one of their hydro systems.

The main objective for the hydro operation is to maximize profit from hydro generation while meeting all physical and judicial constraints on operation. Norway is part of the NordPool electricity markets with well-functioning markets for many different electricity products, the most important being the spot market [1]. The TSO in addition operates short-term regulating markets. It is assumed that the utility is a price taker in all markets.

Because of the complexity of the hydro optimization and scheduling process, the problem is usually divided into different tasks with different planning horizons. Reference [2] gives an overview of the typical planning sequence for a Norwegian hydro producer. The complexity is because decisions today affected future opportunities and because the future is uncertain. The main uncertainties are future prices and inflows to the system. About 50% of the NordPool electricity production is based on hydro and prices are therefore often negatively correlated with local inflows. The optimization and scheduling process include at least the following tasks:

- Medium and long-term price forecasting a few years ahead. Time resolution is typically load periods within the week.

A. Helseth (Ed.): HSCM 2018, *Proceedings of the 6th International Workshop on Hydro Scheduling in Competitive Electricity Markets*, pp. 41–51, 2019.
https://doi.org/10.1007/978-3-030-03311-8_6

- Short-term price forecasting, e.g. hourly some days ahead.
- Snow storage estimation.
- Medium-term inflow forecasting based on snow storage information.
- Short-term inflow forecasting.
- Long-term hydro optimization.
- Medium-term hydro optimization.
- Bidding and short-term hydro scheduling.

The long and medium-term tasks are typically done at least once a week while the short-term tasks are done at least once every day. In this paper the long-term stochastic optimization is for an aggregate hydro model and the medium-term is multi-deterministic for the detailed hydro system. Long and medium-term tasks are taking into account uncertainty in price and inflow while the short-term tasks are done for a deterministic future. This description does not include the very short-term intraday hedging and regulating market activities. The profit from hydro production is a function of all tasks and the importance of each of them depend on the properties of the physical system.

2 Benchmarking Hydro

Benchmarking generally means some form of comparison with other companies or other groups within an organization. The goal of the benchmark is to measure performance compared with others and to identify and motivate for improvements.

Hydro operation involves, as described above, many tasks that are very specific to the system. For example, a very flexible hydro system with good storage capacity cannot be compared with a very poorly regulated system. E.g. long-term price forecasting is not equally important for a poorly regulated system as for a system with high flexibility. For a completely unregulated system the planning is much simpler. The important factors are the availability of the system and ability to make good short-term production (inflow) forecasts.

In Norway, the ownership of some watercourses is split between several owners, typically each owner owns X percent of each reservoir and plant. We will not comment more on the details on how this is arranged but just point to that for such arrangements standard benchmarking can and has been used to compare the whole hydro scheduling process of different companies.

There are two main obstacles to benchmarking the whole hydro scheduling process by comparing with other utilities. Firstly, every hydro system is different as already mention. Secondly, for systems with good storage capacity the strategy for operation might be correct in the long-run but can give poor results in shorter periods because of the uncertainties. The periods need to be very long to filter out chance. It is of course also possible to benchmark parts of the process e.g. the short-term price forecast. This type of approach can be used to track possible improvements from year to year, especially for several of the short-term tasks.

TrønderEnergi already has used a benchmark for their hydro production. The benchmark, shown in Fig. 1, is rather simple and reflects the average price that TrønderEnergi has got for their production relative to the average, non-weighted, price of the year shown in Fig. 2. It shows that the benchmark during the period 2000–2017 on average has improved, but it also shows variations between years that are probably not only due to variations in performance. The benchmark measures the ability to forecasts prices and to produce when the prices are highest but it also includes some non-desirable properties. The benchmark does not penalize flooding, actually more flood or overflow is better because it gives increased flexibility and improved ability to only produce at high prices. Inflow variations between years will also influence flexibility and therefore the possibility to get a good benchmark.

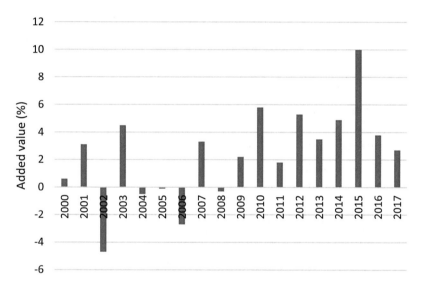

Fig. 1. TrønderEnergi existing benchmark for hydro production. Achieved price divided by yearly average market price.

The benchmarking approach described in this paper is based on a different approach where we have made a computer program that simulates the whole decision process and compared the simulated income from production with the observed income for the historical period. The computer program, which we call the simulator, performs in principle all daily tasks for the whole historical period in sequence. The simulator is programmed in Python and uses APIs to the commercial hydro optimization models for long, medium and short-term optimization.

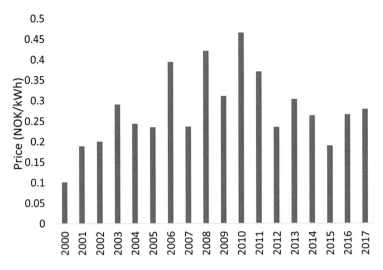

Fig. 2. Average yearly prices for the period 2000–2017.

3 Simulator Components

3.1 Medium and Long-Term Price Forecasting

A medium and long-term price forecast consists of a set of parallel price scenarios for the future, e.g. 208 weeks ahead from current week. There is one price scenario for every weather year that is used to represent inflow uncertainty. The time resolution is load periods within week, e.g. peak hours, night etc. The price forecast is calculated using the EMPS model [3] which is a fundamental based hydro-thermal market optimization and simulation model. Some companies make the forecast themselves and others buy the forecast from consultants. The simulator needs one such price forecast for every week in the historical period. TrønderEnergi had recorded every price forecast they have used for the whole historical period and this is that basis for the analysis.

3.2 Medium and Long-Term Inflow Forecasting

Long-term inflow forecasts are usually based on the assumption that history may repeat itself with equal probability. This is the basis for generating a number, equal the number of years in the history, of parallel inflow scenarios for the future. In this analysis, the weather years 1958 to 2015 is used to represent inflow uncertainty. For the medium-term, i.e. including the coming spring flood period, the forecast based on history is modified to account for the snow storage information. This modification can be done using different methods, a more advanced using HBV models is described in [5]. In our simulator a simpler method based on a form of scaling is used. The scaling factors are user input and changes both the average inflow and the inflow uncertainty for the melting period.

TrønderEnergi had recorded time series for snow storage information and the corresponding inflow scaling factors for every week in the benchmarking period.

3.3   Medium and Long-Term Optimization

The long and medium-term optimization is done in sequence as described in [2]. The long-term optimization is based on stochastic dynamic programming and calculates water values for an aggregate model of the system. These water values are used in a simulation process together with a heuristic to simulate individual reservoir operation. Simulated individual storage levels from the long-term model give target reservoirs to the medium-term model at the end of the medium-term planning horizon.

The medium-term model uses a detailed description of the system in a deterministic optimization for each forecasted scenario. Marginal value of water at the end of the short-term planning horizon (one or two weeks ahead) is calculated by taking the average of each scenario solution. A water value function can be made if the medium-term model is solved for different initial storage levels one or two weeks ahead.

TrønderEnergi also have an alternative SDDP based medium-term optimization model in operation [4]. However, because this model has not been operational from the beginning of benchmarking period and because of computation time, when used in a simulator type approach, the simpler multi-deterministic model has been used in the simulator runs. Except for the mentioned issues, the SDDP based model could have easily been substituted with the medium-term model in the simulator.

3.4   Short-Term Inflow and Price Forecasting

In the analysis, we have chosen not to include uncertainty in the short-term planning period. This is because it makes the implementation of the simulator simpler and because it is assumed that uncertainty in long and medium-term planning is much more important for the specific system that are analyzed. Moreover, records of historical short-term forecasts are not available in this case. Therefore, prices, availability of generators and pump and the inflows to the system are assumed to be known for the whole short-term planning period and equal to the observed values. I.e. the short-term forecast is perfect and known between one and two weeks ahead depending on actual planning day.

We are only considering spot prices in the analysis because income from the other short-term balancing markets only represent a very small part of the income for the benchmarking period.

3.5   Short-Term Optimization

The SHOP model [6] is used for the short-term optimization part. The model uses hourly time resolution and the planning horizon ranging from 9–14 days, depending on the day of the week. The short-term model is run at least once every day. This is because the physical system includes many complicated and state dependent constraints and properties that are not perfectly handled in the optimization. In practice, the model is run every day also because of the real life short-term uncertainties in price and inflows. For our case with perfect forecast the updating is necessary to update to state dependencies and thus to give physically correct connections and to comply with judicial constraints. More on this in Sect. 3.7.

3.6 Availability

The production system is not always available due to outages or planned maintenance. The simulator represents the actual decision problem for each day in the benchmarking period as well as possible and therefore includes information about planned maintenance for the long and medium-term planning tasks and the actual availability of the system for each scheduling day. Planned maintenance does not always correspond to actual availability in operation because of forced outages or change of plans.

3.7 Simulator Logic

The basic structure of the simulator is summarized in Fig. 3. In practice, the daily loop is done twice in parts of the year where there is a judicial discharge constraint that depends of the number of hours with discharge below a certain limit. The first solution is done without the special discharge constraint and is used to calculate an updated discharge constraint for the second run. This is a type of constraint that is not handled internally by the successive linearization and Mixed Integer Programming properties of the short-term optimization model. Simulation results have been verified to ensure that these special constraints also are complied with.

For each week in the benchmarking period 2005-2015

- Input medium and long-term price forecast
- Input medium and long-term inflow forecast and snow storage information
- Input forecasted availability for generators and pump
- Calculate medium-term inflow forecast based on snow storage
- Long-term optimization
- Medium-term optimization

For each day of the week

- — Input observed prices and inflows for the short-term planning period
- — Input actual availability
- — Run short-term optimization
- — Store results for first day
- — Update state dependent model constraints

Fig. 3. Basic simulator structure.

4 The Driva Hydro System

The Driva hydro system is located in the middle of Norway and consists of two reservoirs, one plant with two generators (2 * 75 MW) and one pump, see Fig. 4. Average yearly net production is 630 GWh. The system is, compared to typical Norwegian systems, very simple with respect to the number of plants and reservoirs. However, there are several complicated physical connections and judicial constraints that affect operation. We will not go into all details but just mentioned some of the relations that complicates daily operation:

- Inflow to Gjevilvatnet depends on the actual reservoir level.
- Maximum discharge capacity depends on the reservoir level.
- Minimum discharge constraint depends on reservoir level.
- Discharge constraints depend on the actual discharge the previous day.
- There is a flow constraint downstream the plant that depends on the natural river flow in river Driva.

For the benchmarking results to be precise and trusted all constraints and special physical connections must be complied with. This has been checked for every hour of the benchmarking period. This is the reason why we run the short-term model iteratively for each day as also mentioned previously.

The mathematical models of the hydro system (unit efficiency descriptions, capacities etc.) have been kept constant for the whole benchmarking period. This is a simplification because turbine efficiency is reduced slightly every year and there has been maintenance in the benchmarking period. However, we considered this to have only very minor influence on the results.

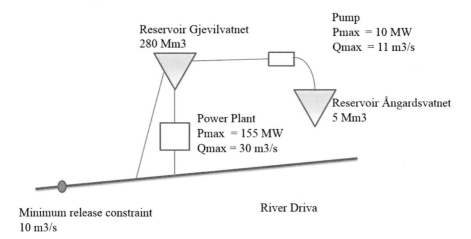

Fig. 4. The Driva hydro system.

5 Case Study

5.1 Benchmarking Using the Simulator

The benchmarking method is basically that we run the simulator for the period 2005–2015 with the information that TrønderEnergi had at each decision stage and compare the simulated results with the observed production, reservoir operation etc. Figure 5 shows simulated and observed reservoir operation for the base case. The base case is actually the results TrønderEnergi would have got if they used the optimization models on a daily basis without manual adjustments. TrønderEnergi have in operation all the optimization models that are utilized in the simulator. The deviation between observed and simulated results may be due to one or more of following:

- Models have not been used or trusted fully throughout the period.
- Operation strategies may have changed during the period, especially risk aversion related to violation of constraints. In the long and medium-term optimization models, risk aversion is indirectly specified by penalties for violation of constraints and the inflow records that are used to inflow uncertainty.
- Modelling errors, the model is not the physical system.

The base case shows what the income could have been, if models where used without user adjustments. The base case results also serve to verify the models and to show that using models would give operation that satisfies all constraints on operation.

The purpose of the benchmark was not only to show that use of optimization tools is useful but also to identify possible improvements and to prioritize what is important and not. To do this the simulator is run for the whole benchmarking period with different input assumptions:

- Different assumptions for the medium and long-term price forecasts, e.g. a forecast based on the forward market, based on fundamental price forecasting models or the actual observed price.
- Time resolution used in the medium and long-term models.
- Type of information coupling between medium and short-term model, constant water values or cuts. Using cuts is in theory the best approach but the individual water values are easier to interpret and more robust to e.g. model inconsistencies.
- With and without snow storage correction of inflow the forecast.

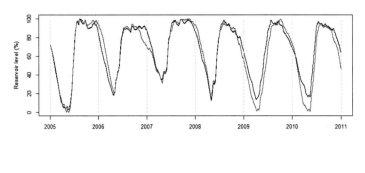

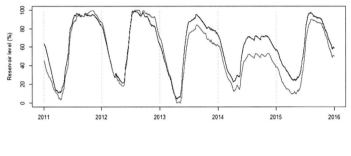

Fig. 5. Observed and simulated (base case) operation of Gjevilvatnet.

6 Results

We will not present all benchmarking results because some of this is company sensitive information. The overall result is that the benchmark gives the importance of the different inputs that are tested for. Maybe not surprisingly, knowing the future price is very important for a system with storage flexibility. Table 1 shows the sum income from hydro production for three cases compared to the observed income. The three cases represent the base case (i.e. blind faith in models and using the information that was available at each time step), a case (Det. Price) where the only difference is that the future price is assumed known for the whole future at every decision stage and case (Const. water values) where the water values that couples the short and medium term models are independent of volume. Inflows are uncertain for all cases.

Table 1 also shows simulated income for all years in the benchmarking period and the sum for whole period. It is very difficult to compare individual years because the storage level at the beginning of the year and at the end of the year may deviate significantly as shown in Fig. 5. The numbers in the Table 1 represent sales value of hydro production at market price, no valuation of storage at the beginning and end of the individual year is included. It is of course possible to put a value to the storage differences based on water values but the precision of such an evaluation will not be robust relative to the size of the changes in income that are evaluated in this project. Chance will also play an important part. The sum results show that knowing the price would have improved the income significantly compared both to the base case (2.3%)

and to the observed values (5.6%). It also shows that using the models, without manual adjustment and with the same parametrization as used in this study, also give significant improvement to the observed income.

Sum simulator production for the benchmarking period is less than observed for all cases (≈150 GWh). This is mainly because of deviation between model efficiencies and real efficiency of the plant and pump. We have check that the deviation is not caused deliberate operation by the model to earn more, e.g. more production above best efficiency (MW) or operation that give more overflow. If model results are corrected for efficiency deviation, a conservative estimate gives that all simulator results should be about 5 mill Euro higher, corresponding to about 2%.

Table 1. Income (million EURO) from hydro production in the Driva system.

	2005	2006	2007	2008	2009	2010
Observed	19.9	24.6	21.3	34.9	18.8	34.8
Base case	18.5	29.6	18.1	36.9	17.5	41.6
Det. price	18.5	35.8	16.0	40.0	13.2	41.8
Const. water values	18.1	29.3	18.4	36.4	17.8	42.0

	2011	2012	2013	2014	2015	Sum
Observed	29.3	24.8	19.5	14.4	11.2	253.5
Base case	25.2	24.3	22.7	14.5	10.3	259.2
Det. price	30.5	23.9	25.9	13.5	8.5	267.6
Const. water values	24.8	24.3	22.0	14.7	10.8	258.6

7 Conclusion

The results from the benchmarking show that TrønderEnergi could have earned about 4.2% more than observed if results from models have been used for the whole period unchanged. This number include 2% correction for use of too low model efficiency. Possible income with perfect price information is estimated to about 7.6% higher than observed. This is of course an unrealistic maximum. The benchmark results also gave useful information about sensitivity to the other inputs that was tested for. The importance of the different inputs is specific to this system.

The computation time for the simulator is very long, about 5–6 h depending on assumptions. For a more complicated water course the computation time would have been even much longer. The main contributing factor is use of unit commitment (MIP) in the short-term optimization model. A medium-term model based on SDDP methodology would contribute to further increase in computation time.

8 Further Work

We plan to further develop the simulator to include functionality that can quantify consequences of errors in short-term price and inflow forecasts. The model would then also give the value of weather forecasts and the hydrological models that's are used to make short-term inflow forecasts. This was not considered to be important for this specific case study but might be very important for other systems.

Further development also includes application of a stochastic short-term model and the ability to evaluate the benefit of such a model. Of course, all of this would increase computation times even further

References

1. https://www.nordpoolgroup.com/Market-data1/#/nordic/chart
2. Fosso, O.B., Gjelsvik, A., Haugstad, A., Mo, B., Wangensteen, I.: Generation scheduling in a deregulated system: the Norwegian case. IEEE Trans. Power Syst. **14**(1), 75–81 (1999)
3. Wolfgang, O., Haugstad, A., Mo, B., Gjelsvik, A., Wangensteen, I., Doorman, G.: Hydro reservoir handling in Norway before and after deregulation. Energy **34**, 1642–1651 (2009)
4. Gjelsvik, A., Mo, B., Haugstad, A.: Long- and medium-term operations planning and stochastic modelling in hydro-dominated power systems based on stochastic dual dynamic programming. In: Rebennack, S., Paradalos, P.M., Pereira, M.V.F., Iliadis, N.A. (eds.) Handbook of Power Systems I, pp. 33–55. Springer, Heidelberg (2010). ISBN: 978-3-642-02492-4
5. Lindstrøm, G., Johansson, B., Persson, M., Gardelin, M., Bergstrøm, S.: Development and test of the distributed HBV-96 hydrological model. J. Hydrol. **201**, 272–288 (1997)
6. Fosso, O.B., Belsnes, M.M.: Short-term hydro scheduling in a liberalized power system. In: 2004 International Conference on Power System Technology - POWERCON, vol 2, pp. 1321–1326 (2004)

Optimal Pricing of Production Changes in Cascaded River Systems with Limited Storage

Hans Ole Riddervold[1(✉)], Hans Ivar Skjelbred[2], Jiehong Kong[2],
Ole Løseth Elvetun[3], and Magnus Korpås[4]

[1] Norsk Hydro and NTNU, Trondheim, Norway
hans.o.riddervold@ntnu.no
[2] SINTEF Energy Research, Trondheim, Norway
[3] Nextbridge Analytics aS, Oslo, Norway
[4] NTNU, Trondheim, Norway

Abstract. Optimization of hydroelectric power production is often executed for river systems consisting of several powerplants and reservoirs located in the same region. For hydropower stations located along the same river, the release from upstream reservoirs ends up as inflows to downstream stations. Calculating marginal cost for a string of powerplants with limited reservoir capacity between them, requires a new approach compared to heuristically calculating marginal cost for single plants in well-regulated hydrological systems. A new method, using marginal cost curves for individual powerplants to generate an overall marginal cost curve for interlinked power stations has been developed. Results based on a real-world case study demonstrate the advantage of the proposed method in terms of solution quality, in addition to significant insight into how optimal load distribution should be executed in daily operations.

Keywords: Heuristic algorithms · Hydroelectric power generation
Cascaded river systems

1 Introduction

In the planning process for production of hydroelectric power, the optimal solution associated with predicted prices and inflows can be used to create bids for the day-ahead spot market and generate production schedules [1]. Deviation from the original production scheduling, typically created 12–36 h prior to actual production hour, is more frequent with increasing activity in the intraday-market and more volatility imposed by intermittent power production.

For a power producer, it can be tempting to optimize all power stations and reservoirs located in the same price area in one common model. A motivation for this could be distribution of obligations in the spot and reserve markets, and/or for financial hedging purposes [2]. For practical purposes, and to reduce calculation time, optimization is often carried out on an aggregation level where hydraulically coupled reservoirs and power stations are modeled together. For river systems with large reservoir capacity between power stations, the interdependency between production in

© Springer Nature Switzerland AG 2019
A. Helseth (Ed.): HSCM 2018, *Proceedings of the 6th International Workshop on Hydro Scheduling in Competitive Electricity Markets*, pp. 52–61, 2019.
https://doi.org/10.1007/978-3-030-03311-8_7

the individual plants could be more important for long-term maneuvering than for short-term bidding. However, for cascaded river systems with limited storage capacity between plants, hereafter referred to as linked river systems, this is often the opposite case since production in a downstream plant is a direct result of upstream production.

The existing method used for comparison in the case study investigated in this paper is based on the "Single Plant Model" [3]. For application of the Single Plant Model, Hveding's conjecture [4] states that "in the case of many independent hydropower plants with one limited reservoir each, assuming perfect maneuverability of reservoirs, but plant-specific inflows, the plants can be regarded as a single aggregate plant and the reservoirs can be regarded as a single aggregate reservoir when finding the social optimal solution for operating the hydropower system". This model uses basic principles for energy calculations considering, head-loss, generator- and turbine efficiency to generate a combined production/waterflow relationship for the single plant, and thereby associate marginal cost to different levels of operation.

Other methods for scheduling and/or coordinated control in cascaded river systems have been described [5–7]. These approaches apply different optimization techniques, and are often tailored for individual river systems. They are not necessarily primarily designed for bidding in the spot-, balancing- and intraday market, but could represent an alternative approach to the method described in this paper. An important criterion in relation to finding an applicable method to be used in the bidding process, is the time used to generate bids.

A method for heuristically calculating the marginal cost for all the operating points of a power plant, covering the entire working area for the plant and including all the physical limitations and reserve obligations in other markets has been presented by SINTEF [8]. For a hydrological system with significant storage, the method has demonstrated to be computationally efficient.

In this paper, we present a short-term scheduling method for heuristically calculating the marginal cost in linked river systems where storage capacity between plants is limited.

The method has further been investigated on a large Norwegian river system consisting of five linked powerplants with varying degree of interim storage capacity. The results from the calculation have been used to generate dynamic bids for rapid response to opportunities in the intraday and balancing market as market prices, inflows, and other physical parameters in the river system change.

2 Problem Description

When heuristically calculating the marginal cost for plant production, the optimum production for one plant can be associated with a different waterflow than for another plant in the river system. Results from methods developed for hydrological systems with significant storage capacity between plants can therefore not be directly applied.

The main challenge associated with computing marginal cost for linked power plants to be used for bidding in the spot, balancing and/or intraday market, is that the waterflow for the power stations must be in balance at all time-steps.

2.1   Existing Method Used for Calculating Marginal Cost for Linked Power Plants

Water values as marginal cost for hydropower generation is a widespread means of assigning monetary values to the available water resources. The water value can be defined as the future expected value of the stored marginal kWh of water, i.e. its alternative cost [9, 10]. The water value for a power station is typically given by a seasonal model, and referred to the optimal point of operation (Q*) for the power plant.

$$Q^* = \arg\max_{Q} \frac{P(Q)}{Q} \tag{1}$$

From basic economic theory, the marginal cost for one operating point is the change in the opportunity cost of water (C) involved as a result of an infinitesimally small increase in the discharge of the units (P), which is expressed as:

$$mc = \frac{\partial C}{\partial P} \tag{2}$$

Combining the water value as reference for marginal cost at optimal point of operation with (1), the piecewise linear marginal cost (€/MWh) for changing production from production i to j (MW) is given by (3)

$$mc_{ij} = \frac{\Delta Q_{ij}}{\Delta P_{ij}} * \alpha * WV^* \tag{3}$$

Where WV* is the water value (€/MWh) at optimal point of operation and α is given by ∂P divided by ∂Q at optimum (Q*). For discrete mc calculations, α is fixed to a value such that mc is equal to WV* at the point of operation where the highest production relative to the water consumption is defined. This method is used when the existing method calculates marginal cost for discrete change between predefined levels of production.

Assuming that a plant can operate independently, and a water value referred optimal production of 30 €/MWh, Table 1 illustrates how marginal cost can be calculated using (3). Plant 1 would in this case produce 100 MW (35 m3/s) at a market price of 30 €/MWh.

Table 1. Marginal cost for "independent" power station using existing method

Plant 1 P [MW]	Plant 1 Q [m3/s]	P/Q	$\frac{\Delta Q_{ij}}{\Delta P_{ij}}$	mc_{ij}
70	25	2.80		
100	**35 (Q*)**	**2.86**	**0.33**	**30.0**
140	50	2.80	0.38	33.8
200	75	2.67	0.42	37.5

In this example, we are calculating the marginal cost of increasing production from level i to j. The marginal cost for the initial P/Q level is therefore omitted.

If we introduce a second plant in the example with optimal production at another waterflow, and these plants are linked with limited or no intermediate storage capacity, a common waterflow must be chosen where one or both power stations must deviate from the plants optimal point of operation (Q*) to avoid flooding past one of the plants. The existing approach is to aggregate the production (MW) for the plants at the same waterflow (m3/s) to create a common production/waterflow relationship for the two plants.

Even though some of these river systems originally where designed to have common optimal waterflow, gradual plant upgrades and market developments affecting production patterns, might lead to the need of making tradeoffs between optimum production in the different plants.

2.2 Proposed Method

The existing method described in Sect. 2.1 defines a common production-waterflow curve for a linked river system. The main weakness of generating a curve based on this method concerns the dynamics that are associated with modeling of several power stations in a linked river system. One plant could consist of several generators where some are shut down for maintenance. There could also be temporary load restrictions, concessional requirements, or local inflow effecting operations. This would require a continuous update of the combined production-waterflow curve. It would also require maintenance of a model which is not representing the physical power system. Finally, when distributing load requirements, a separate model or optimization must be run to allocate production to the correct generators.

An improved method is described in the two following sections. The first section gives a general description of the best profit method, while the second describes how the method can be used for linked river systems.

2.2.1 Heuristics, Best Profit

For completeness, we include a description of the way marginal cost curves are created by the best profit functionality in the Short-term Hydro Optimization Program (SHOP). SHOP is a software tool for optimal short-term hydropower scheduling developed by SINTEF Energy Research [1]. Interested readers can find more details about the best profit functionality in [8]. We assume that the water value and gross head for each plant is given. In real-world operation, the head loss in the main tunnel and the penstock that unit i connects to should not be neglected. It can be represented as a quadratic equation of the total flow going through the main tunnel/penstock. The net head, and therefore, is calculated as:

$$NH_{ist} = GH_{st}$$

$$- \alpha_{main} \cdot \left(\sum_{i \in I_{s,main}} q_{ist} \right)^2 - \alpha_{pen} \cdot \left(\sum_{i \in I_{s,pen}} q_{ist} \right)^2 \qquad (4)$$

$$i \in I_s, s \in S, t \in T$$

where:

I_s	Set of units in plant s
$I_{s,main}$	Set of units that connect to main tunnel in plant s.
$I_{s,pen}$	Set of units that connect to penstock pen in plant s.
NH_{ist}	Net head of unit i in plant s at period t (m).
α_{main}	Loss factor for main tunnel.
α_{pen}	Loss factor for penstock pen.
q_{ist}	Flow going through unit i in plant s in period t (m^3/s)

For a generating unit i in plant s, the power production, in (5), depends on the net head and the flow going through that unit. It also relies on the generator efficiency and head-dependent turbine efficiency.

$$mw_{ist} = 0.001$$

$$\cdot \eta_i^{GEN}(mw_{ist}) \cdot \eta_i^{TURB}(q_{ist}, NH_{ist}) \cdot G \cdot NH_{ist} \cdot q_{ist} \qquad (5)$$

$$i \in I_s, s \in S, t \in T.$$

where:

mw_{ist}	Power produced by unit i in plant s in period t (MW).
η_i^{GEN}	Generator efficiency of unit i, which is interpolated on the basis of production mw_{ist}.
η_i^{TURB}	Turbine efficiency of unit i, which is interpolated on the basis of flow q_{ist} and net head NH_{ist}.
G	Gravity value, default setting is 9.81 (m/s^2)

Based on (4) and (5), if the discharge for each unit is given (i.e. one possible operating point for the plant), we can precisely calculate the corresponding production, taking the head loss into consideration. This transformation from the flow discharge to the power generation is implicitly done by the functionality in SHOP.

For a given operating point p in one specific unit combination c, the generation cost for this point is the opportunity cost of the water used. In the previous section, we have presented how the hourly water cost is defined, and how the production can be accurately obtained when the discharge of the units is decided, in (5). Therefore, we denote the average cost for this operating point by

$$ac^p_{cst} = \frac{3600 \cdot WC_{st} \cdot \sum_{i \in I_c} q^p_{ist}}{\sum_{i \in I_c} mw^p_{ist}} \tag{6}$$

$$p \in P_c, c \in C, s \in S, t \in T.$$

where:

C Set of unit combinations.

I_c Set of units in unit combination c.

P_c Set of operating points in unit combination c.

ac^p_{cst} Average cost for the operating point p in unit combination c in plant s in period t (€/MWh).

In economics, marginal cost is the change in the opportunity cost that arises when the quantity produced has an increment by one unit. In contrast to the transformation from discharge to production, it is much more complicated to find the discharge by a given production. In addition, the power produced is infinitely divisible. Therefore, we find the marginal cost by increasing the discharge by a small amount, expressed as

$$mc^p_{cst}$$

$$= \frac{3600 \cdot WC_{st} \cdot \sum_{i \in I_c}(q^p_{ist} + \Delta q^p_{ist}) - 3600 \cdot WC_{st} \cdot \sum_{i \in I_c} q^p_{ist}}{\sum_{i \in I_c}(\widetilde{mw}^p_{ist}) - \sum_{i \in I_c} mw^p_{ist}} \tag{7}$$

$$p \in P_c, c \in C, s \in S, t \in T.$$

where:

mc^p_{cst} Marginal cost for the operating point p in unit combination c in plant s in period t (€/MWh).

Δq^p_{ist} A small increment in the discharge of unit i in unit combination c in plant s in period t, $\Delta q^p_{ist} = 0.001 \cdot \frac{q_{ist}}{\sum_{i \in I_c} q_{ist}}$, (m³/s).

$\widetilde{mw}^p_{ist}$ Power produced by unit i in plant s in period t when there is a small increment in the discharge of the units (MW).

After calculating the marginal cost for a large number of combinations of flows in the running units, we can find the optimal production distribution and the corresponding marginal cost curve. For each production level, the optimal distribution is the one resulting in the lowest discharge. This ensures that the most efficient units will always be used first.

Best profit curves generally contain information about the marginal cost of each production level, the optimal production distribution between the running units and at what price it is optimal to switch between unit combinations. Example of a Best profit curve can be found in Fig. 2. In this paper, it is assumed that the combination of running units at each plant is given. This means that the best profit curve only has to contain the information about marginal costs and optimal production distribution, as described above.

2.2.2   Best Profit Customized for Linked River Systems

To be able to apply the best profit method for linked river systems, output from the model must include a marginal cost [€/MWh]/waterflow [m3/s] relationship in addition to the marginal cost [€/MWh]/production [MW] relationship that is already produced by the existing models. Several requirements must be met for the proposed method to be applicable.

To keep focus on the main principles of the best profit method compared to the existing method, a requirement in this analysis is set that all generators in the river system must run.

A water-value is normally estimated for each power station in the river system, and the value is defined as the marginal cost of an incremental increase of production from the optimal point of operation for the specific power station. This leaves us with a challenge related to defining which water-value should be used for the aggregated power station. In the best profit calculations presented in this paper, one common water-value is used for all plants, and is estimated as a weighted average of the water-values for the power stations at the combined plants optimal production.

If we assume there is no intermediate natural inflows between stations, calculating the aggregated marginal cost value for the linked river system requires that plant specific marginal costs are selected for identical waterflows for each power station. For power station 1 to 5, $q1 = \cdots = q5$. The method can be extended to handle inflows between stations.

To accommodate for the difference in output effect for the power stations, the marginal cost for each plant must be weighted according to the relative production of the plant at the selected waterflow to generate the overall marginal cost at a selected point of operation.

Each bidding point will have a unique waterflow and accumulated production associated with it. We can therefore select the price that should be used for bidding of production (MW) to the spot, intraday or balancing market. For the intraday and balancing market, the bidding price would typically be the price associated with deviating from the current point of operation.

3   Case Study

The river system investigated in this paper consists of five plants with very limited intermediate storage (Fig. 1). The main reservoir which is located upstream all power plants has a degree of regulation of approx. 1.5, meaning that yearly inflow is 1.5 times reservoir capacity. To produce the yearly inflow, the power stations must operate at full capacity approximately 70% of the available hours in a year. The time used from when water is released from the main reservoir until it reaches the lower reservoir is short and is disregarded in this analysis. The lower reservoir has sufficient storage capacity, and the linked river system can be assumed to operate independently of the water content in this reservoir. To reflect the alternative value of production, a water value of 31 €/MWh from the main upstream reservoir is used as basis.

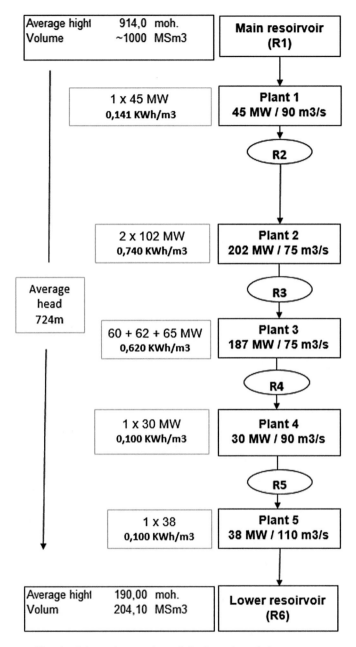

Fig. 1. Schematic overview of the investigated river system

The best profit values have been selected for production in an hour where market price is equal to the water-value (31 €/MWh). At this price, all power stations are in operation. To ensure that all generators are running, q_min = 60 m3/s and q_max = 75 m3/s in further marginal cost calculations.

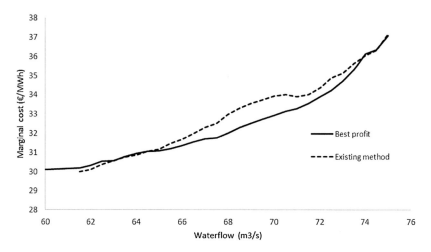

Fig. 2. Aggregated marginal cost curves for the linked river system

Figure 2 illustrates how the best profit curve compares to the existing method described in Sect. 2.1. The graph shows that the result coincides well for production in the lower and upper range of the waterflow area. However, there is a deviation in production in the mid-range of water flows.

The observed deviation reveals a true benefit of using a dynamic best profit curves. Plant 3 is one of the larger power stations in the river system consist of 3 generators. This plant has been through several upgrades during the last years. These generators have different characteristics, and how these generators are uploaded will have significant impact on the plants total efficiency. In the existing method, these generators are uploaded in steps defined by a relatively simple algorithm, whilst the best profit utilize the complimentary characteristics of the generator to ensure optimal distribution of load for all represented waterflows. This results in a relatively low loss of efficiency for the plant for waterflows in the range from 60-67 m3/s compared to the other plants.

4   Conclusion

It has been demonstrated that the best profit method can be used to generate real-time marginal cost curves for linked river systems. These results can readily be used for bidding to the spot-, balancing and/or intraday market. Further, results from the best profit give significant insight into how optimal load distribution for linked river system should be executed in daily operations. Often, real-time regulation of complex linked river systems is carried out by SCADA-systems with limited user interaction from production planners handling the commercial process. Having a quick and robust method like best profit available, the traditional and often static approach to operation of linked river systems can continuously be challenged. This will create additional values for the power producers, and ensure that pricing toward a gradually more complex market is as correct as possible.

In this paper, we calculate the marginal cost for a linked river system where all generators are in operation. This is not a limitation that applies for the method in general. A relatively trivial expansion is to investigate the best profit value when one or more generators are out for maintenance. This can be done by investigating the best profit values for an area of operation where the available generators are running. The existing existing-method however, has considerable challenges in handling these situations. For further analysis, it will also be of interest to investigate production behavior in ranges where different generators in the linked river system will be turned on and off. Defining more correctly the link between water values and the use in the best profit method, particularly to incorporate the coupling with cuts [11], will also be an issue for further improvement.

References

1. Fosso, O.B., Belsnes, M.M.: Short-term hydro scheduling in a liberalized power system. In: Proceedings of 2004 International Conference on Power System Technology - POWERCON, vols. 1 and 2, pp. 1321–1326 (2004)
2. Mo, B., Gjelsvik, A., Grundt, A.: Integrated risk management of hydropower scheduling and contract management. IEEE Trans. Power Syst. 16, 216–221 (2001)
3. Hveding, V.: Digital simulation techniques in power system planning. Econ. Plan. 8(1), 118–139 (1968)
4. Forsund, F.: Hydropower Economics, 2nd edn. International Series in Operations Research & Management Science), vol. 112, pp. 71–76. Springer US, Boston (2015)
5. Hamann, A., Hug, G., Rosinski, S.: Real-time optimization of the mid-columbia hydropower system. IEEE Trans. Power Syst. 32(1), 157–165 (2017)
6. Nilsson, O., Sjelvgren, D.: Mixed-integer programming applied to Short term planning of a hydro-thermal system. IEEE Trans. Power Syst. 11(1), 281–286 (1996)
7. Aasgard, E.K., Andersen, G.S., Fleten, S.E., et al.: Evaluating a stochastic-programming-based bidding model for a multireservoir system. IEEE Trans. Power Syst. 29(4), 1748–1757 (2014)
8. Skjelbred, H.I., Kong, J., Larsen, T.J., Kristiansen, F.: Operational use of marginal cost curves for hydropower plants as decision support in real-time balancing markets. In: 2017 14th International Conference on the European Energy Market (EEM), pp. 1–6 (2017). ISSN 2165-4093
9. Wolfgang, O., Haugstad, A., Mo, B., Gjelsvik, A., Wangensteen, I., Doorman, G.: Hydro reservoir handling in Norway before and after deregulation. Energy 34(10), 1642–1651 (2009)
10. Brovold, S.H., Skar, C., Fosso, O.B.: Implementing hydropower scheduling in a european expansion planning model. Energy Procedia 58, 117–122 (2014)
11. Gjelsvik, A., Mo, B., Haugstad, A.: Long- and medium-term operations planning and stochastic modelling in hydro-dominated power systems based on stochastic dual dynamic programming. In: Handbook of Power Systems I, pp. 33–56, Springer, Heidelberg (2010)

Modelling Tunnel Network Flow and Minimum Pressure Height in Short-Term Hydropower Scheduling

Per Aaslid[1]([⊠]), Hans Ivar Skjelbred[1], and Sigri Scott Bale[2]

[1] SINTEF Energy Research, Sem Sælands vei 11, 7034 Trondheim, Norway
Per.Aaslid@sintef.no
[2] Hydro Energy AS, Drammensveien 364, 0283 Oslo, Norway

Abstract. The paper proposes a method for modelling tunnel network flow between reservoirs and creeks above hydro power plants in short term hydro optimization. A method for handling pressure constraints in nodes in the tunnel network is also included. The method is applied on a plant below a reservoir and a creek, with a rigorous minimum pressure constraint in a tunnel. A comparison of the presented method with a manual adjustment method for handling the minimum pressure constraint shows a 3.3% increase in objective value of the original total sale.

Keywords: Hydropower scheduling · Optimization · Network hydraulics

Nomenclature

Sets and indices

$I = \{1, \ldots, i\}$	Set of nodes in the junction network
$J_i \in I$	Set of nodes adjacent to node i
R	Set of reservoirs in the junction network, where r_i is in node i
C	Set of creeks in the junction network, where c_i is in node i
$E_{xy} = \{(x, i_1), \ldots, (i_n, y)\}$	Set of edges between node x and y. (Used to traverse between nodes.)

Parameters

Q_{ij}	Tunnel flow from node i to j (m^3/s)
Q_i^{CR}	Creek inflow in node $i (m^3/s)$
H_i	Head in node $i(m)$
H_i^{CR}	Creek height in node $i(m)$
H_i^{MIN}	Minimum pressure in node $i(m)$
V_i	Storage in reservoir in node $i(10^6 m^3)$
α_{ij}	Tunnel loss factor between node i and $j(s^2/m^5)$

© Springer Nature Switzerland AG 2019
A. Helseth (Ed.): HSCM 2018, *Proceedings of the 6th International Workshop on Hydro Scheduling in Competitive Electricity Markets*, pp. 62–68, 2019.
https://doi.org/10.1007/978-3-030-03311-8_8

Variables

q_{ij} Tunnel flow from node i to $j(m^3/s)$
q_i^S Spillage from creek in node i (m^3/s)
h_i Head in node i (m)
v_i Reservoir storage in node i $(10^6 m^3)$

1   Introduction

In this paper, we describe a method for modelling tunnel network flow and pressure above a plant in SHOP (Short-term Hydro Optimization Program), which is a software tool based on Successive Linear Programming (SLP), and is in operational use by several hydro power producers [1]. Several Norwegian water courses have high head plants and complex water courses with several reservoirs and creeks above the plant. Many of these have increased the production capacity since the original plant was constructed, and are subject to rigorous pressure constraints due to high tunnel losses. These topologies demand detailed modelling of the flow and pressure in the tunnel networks to produce applicable production plans considering pressure constraints in the tunnels.

[2] proposes a method for modelling junction in short-term hydro optimization limited to one junction and two reservoirs. Moreover, [3] suggest a method for solving the minimum pressure problem for a specific plant, but the method is limited to two plants and one creek intake, and the minimum pressure constraint is located where the creek flow enters the tunnel. The method suggested in this paper is more general since it allows combining an arbitrary number of reservoirs and creeks in the same tunnel network, and minimum pressure constraints could be located anywhere in the network.

2   Method

A junction network is a set of nodes and edges where the nodes represents either a junction, reservoir or creek, and the edges represent the tunnels connecting the nodes as shown in Fig. 1. Each tunnel (edge) is associated with a loss coefficient α shown as a number in the figure. In a junction network, reservoirs, creeks and plants will always represent end nodes in the network, while junctions connects several tunnels in the network. In this paper the number of plants in a network are limited to one, and the power- and head loss between plant and junction are treated separately from this method.

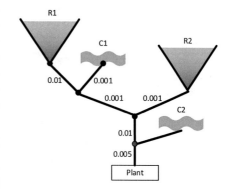

Fig. 1. Example topology

We assume that all tunnels are filled with water and that the head loss is quadratic as defined in the Darcy-Weisbach equation [4]. Given the node pressure h_i and the tunnel flow q_{ij}, the tunnel flows are given by the mass and pressure balance equations in (1) and (2). The mass balance ensures that the total node inflow balances the outflow, while the pressure balance ensures that the head loss for a tunnel is given by $\alpha q|q|$ where α is a loss coefficient for a given tunnel.

$$\sum_{j \in J_i} q_{ij} = 0, i \in I \tag{1}$$

$$h_i - \alpha_{ij} q_{ij} |q_{ij}| = h_j, i \in I, j \in J_i \tag{2}$$

However, the pressure balance Eq. (2) has to be linearized (3) to fit into the SLP scheme. Q_{ij} is given by the tunnel flow from previous iteration. To determine Q_{ij} for the first iteration, we assume maximum plant production, and distribute the tunnel flow proportional with the reservoir and creek inflow and the reservoir size above the respective tunnel.

$$h_i - h_j = \alpha_{ij} q_{ij} |Q_{ij}|, i \in I, j \in J_i \tag{3}$$

The pressure balance in (3) is only applicable between two adjacent nodes. For two arbitrary end nodes in the same network, the linearized pressure balance is given by (4).

$$h_i - h_j = \sum_{(x,y) \in E_{ij}} \alpha_{xy} q_{xy} |Q_{xy}|, (i,j) \in I \tag{4}$$

Finally, the reservoir head is a function of the volume. Each reservoir has a volume balance restriction per time step, where the volume for next time step is expressed as the sum of current volume and net inflow. The reservoir head is given by linearizing the reservoir head as a function of volume as shown in (5).

$$h_i = H_i + \frac{\partial h_i}{\partial v_i}(v_i - V_i), i \in R \tag{5}$$

The resulting pressure balance between two reservoirs are given by inserting (5) for h_i and h_j into (4).

Creek nodes will have a fixed inflow until the pressure in creek tunnel exceeds the creek height. Excessive flow will then be spilled. The linearized pressure balance between a creek and a reservoir are given by (6).

$$H_i^{CR} - h_j \geq \sum_{x,y \in E_{ij}} \alpha_{xy} q_{xy} |Q_{xy}|, i \in C, j \in R \tag{6}$$

The resulting pressure balance between creek and reservoir is obtained by inserting (5) for h_j in (6). Furthermore, the mass balance for creek inflow, tunnel flow and spillage is given by (7), where there is an overflow cost associated with q_{ij}^S.

$$q_{ij} + q_{ij}^S = Q_i^{CR}, i \in C, j \in J_i \tag{7}$$

Finally, head loss is compensated by building a piecewise linear loss function for each tunnel, and distributing it on generator level using the same method as described for pumps in [5] Eq. (20).

For a junction network with n junction nodes and 3 tunnels from each junction node, the total number of creeks and reservoirs is $(n + 1)$. The pressure balance equations are only used to describe the pressure balance between two reservoirs, or a creek and a reservoir. The theoretic maximum number of pressure balance equations for a network is $\frac{n}{2}(n + 1)$. Moreover, it is sufficient that each tunnel in the network is covered by one pressure balance equation.

In order to describe the method for selecting which pressure balance to incorporate in the SLP-scheme, we will first define best reservoir R_i^{BEST}. For an arbitrary node i in a junction network, node i's best reservoir is the reservoir that has the lowest tunnel loss between the reservoir and the node as defined in (8).

$$R_i^{BEST} = \arg \min_{j \in R} \sum_{x,y \in E_{ij}} \alpha_{xy}, i \in I \tag{8}$$

Based on this definition, we select which pressure balances to build as follows

1. For a reservoir or a creek in the network in node i, find R_i^{BEST}.
2. Build pressure balance between node i and R_i^{BEST} using (4) and (5) for reservoirs and (5) and (6) for creeks.
3. Repeat 1 and 2 for all creeks and reservoirs not yet included in any pressure balance.
4. For edges (i, j) not included in any pressure balance, build pressure balance between R_i^{BEST} and R_j^{BEST}.

Given a minimum pressure restriction in node j, the restriction is implemented as shown in (9).

$$h_i - \sum_{x,y \in E_{ij}} \alpha_{xy} q_{xy} |Q_{xy}| \geq H_j^{MIN}, i = R_j^{BEST}, j \in I \tag{9}$$

To illustrate the procedure, we will use the example topology in Fig. 1. We start with C1, where the best reservoir based on (9) is R2, and build the pressure balance using (5) and (6). Then we build the pressure balance from C2 to the best reservoir R2. Best reservoir for R1 is R2, and we use (4) and (5) to build the pressure balance. Finally, there is minimum pressure restriction at the point above the plant which is built from that point towards the best reservoir which is R2 using (9).

3   Results

The method proposed in this paper is implemented on one of Hydro Energy's plants where
the topology is shown in Fig. 2. There is a settling basin marked with red below the junction
that has a minimum pressure constraint at
991 ms. Moreover, the plant is subject to high
tunnel loss, around 20 m between the reservoir
and the settling basin at maximum production,
and the total head loss is even larger. The reservoir is emptied in less than 10 days at maximum
production if the inflow is low.

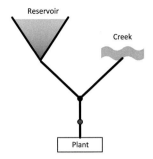

Fig. 2. Test case system

The presented results show 14 days optimization, with 3 iterations to find the
optimal unit commitment and 5 iterations with locked unit commitment to find optimal
production. The largest difference in production between the two last iterations is
56 MW (22% of maximum capacity) at one time step, but on average only 0.68 MW
(0.27%) for the entire optimization period. This shows some potential for improvement
in the convergence properties.

The water value used for the reservoir in this case is fixed. Water value is the
marginal cost of the water calculated in a long- or mid-term model. The value represents the value of the remaining water in the end of the optimization period. We also
assume that we have a perfect prediction of the price. The price in the case is a
historical prognosis. The first chart in Fig. 3 shows that the price is above the water
value for almost the entire optimization period. The inflow to the reservoir and creek is
on average less than 5% of the production.

Case 1 shows the result when optimizing without the minimum pressure constraint.
The production is close to maximum for the entire period resulting in almost empty
reservoir at the end. Furthermore, the pressure constraint is severely violated, hence the
production plan is not applicable for operational use.

To mitigate the violation of the pressure constraint, a manual adjustment method
similar to the one used by the producer have been used. After optimizing, a maximum
discharge constraint is added to the problem if the pressure restriction is violated. The
procedure is repeated several times until we have a feasible plan. As seen from the
results in case 2, this method does not capture the price variations and the result is a
quite flat production for the entire period.

Finally, case 3 shows the production with minimum pressure restriction in SHOP,
and we see that the production in the beginning of the period is moderate such that the
reservoir level is high enough to allow high production when the prices are at its
highest between hour 216 and 264.

Comparing case 2 and 3 shows 0.9% reduction in sale and 4.2% increase in
reservoir end value. That is a 3.3% increase in value of the sale from case 2 to 3.

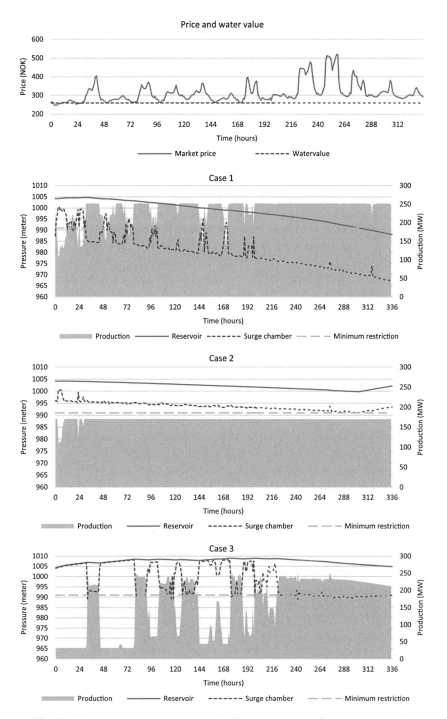

Fig. 3. Price and water value, and production and pressure for case 1, 2 and 3.

There are some minor violations of the pressure constraint due to numerical inaccuracy from linearization and the iterative procedure. These could be compensated for by setting a more conservative pressure restriction. Increasing the pressure constraint with 1 m reduces the objective with 0.41%, which indicates that the added value from the new model still is significant after compensating for pressure constraint violation.

4 Conclusion

The method implemented in this paper enables modelling of complex tunnel networks in short-term hydro power scheduling, and produces a feasible production plan without violating rigorous pressure constraints in tunnel networks.

For the case presented in this paper, the proposed method increases the value of sale and remaining water with 3.3% of the total sale for a two weeks optimization compared to a manual adjustment procedure. The results show that incorporating the pressure constraint in the SLP scheme utilizes the high prices late in the optimization period far better than the manual method that tends to produce almost constantly the entire period.

Other situations where the model could possibly lead to improved schedules is when the reservoir level is low and the creek inflow is high. In this situation, high production is possible since the creek inflow will reduce the discharge and head loss in the tunnel from the reservoir to the junction. When the reservoir level is high, the minimum pressure constraint will not be binding, hence the improved model will have less impact.

A further improvement of the solution would be to include the minimum pressure constraint in the mid-term model that is generating the water values.

One idea for better convergence properties is to introduce segments for the minimum pressure constraint as described in [3] for a more precise representation of the quadratic loss characteristics of the tunnels.

References

1. Fosso, O.B., Belsnes, M.M.: Short-term hydro scheduling in a liberalized power system. In: International Conference on Power System Technology (2004)
2. Belsnes, M.M., Roynstrand, J., Fosso, O.B.: Handling state dependent nonlinear tunnel flows in short-term hydropower scheduling. In: International Conference on Power System Technology (2004)
3. Dorn, F.B., et al.: Modelling minimum pressure height in short-term hydropower production planning. Energy Proc. **87**, 69–76 (2016)
4. Weisbach, J.L.: Lehrbuch der ingenieur-und maschinen-mechanik: Statik der bauwerke and mechanik der umtriebsmaschinen. Vieweg (1848)
5. Kong, J., Skjelbred, H.I., Abgottspon, H.: Short-term hydro scheduling of a variable speed pumped storage hydropower plant considering head loss in a shared penstock. In: Proceeding of 29th IAHR Symposium on Hydraulic Machinery and Systems, Kyoto, Japan (2018)

Implied Efficiency Curves from Analysis
of Operational Patterns

Sebastian Brelin[1][✉], Morten A. Lien[1], Stein-Erik Fleten[1],
Jussi Keppo[2], and Alois Pichler[3]

[1] Norwegian University of Science and Technology,
7491 Trondheim, Norway
stein-erik.fleten@ntnu.no
[2] NUS Business School, National University of Singapore,
Singapore 119245, Singapore
[3] Chemnitz University of Technology, 09111 Chemnitz, Germany

Abstract. A reservoir manager at a hydropower plant has to decide whether to release water in order to produce electricity, and the level at which to produce. These production levels have different efficiencies as well as other related technical aspects. Often, the plant will produce at the best efficiency point, i.e. release water at a rate that produces the highest amount of electricity per unit of water. We apply a structural estimation approach to a hydropower plant in the Norwegian electricity price zone NO5, in order to discover the managers' preferences related to the different production levels. We use time series models in order to replicate the managers' expectations of future conditions. The results show a greater willingness of the manager to produce at levels below than above the best efficiency point, which we argue is mainly due to the increased level of cavitation. They also imply that the reservoir managers' preferences have changed over time, showing an increased willingness to produce at production levels both above and below the most efficient level.

Keywords: Hydropower · Structural estimation
Restructured electricity markets · Mid-term scheduling

1 Introduction

We study operational patterns from hydropower generation, for a price-taking producer participating in a well-functioning market. A hydropower producer faces the challenge of choosing the production level in order to make the end benefit as high as possible. A turbine has a specific efficiency curve that incentivizes production at the best efficiency point (BEP). However, producing at a higher level when prices are high might result in a higher total revenue despite the loss of efficiency. Likewise, producing at a lower level in order to store water when anticipating rising prices might also result in higher revenues. Further, the production efficiency curve changes over time due to damages to the turbine. In addition, some levels of production entail certain problems, such as increased maintenance costs and cavitation. The production policy that the operator uses might hide economic preferences that are not explained by the

© Springer Nature Switzerland AG 2019
A. Helseth (Ed.): HSCM 2018, *Proceedings of the 6th International Workshop
on Hydro Scheduling in Competitive Electricity Markets*, pp. 69–75, 2019.
https://doi.org/10.1007/978-3-030-03311-8_9

mechanical loss of efficiency itself. Discovering these preferences can be valuable in the analysis of a hydropower plant, both for outsiders and for the reservoir managers themselves. The results can be used in internal discussions of whether they reflect the intended operational policy of the plant.

This paper contributes by providing an approach to backing out the efficiency curve, the efficiency as a function of unit load, based on a time series of hydropower releases. The perspective is that of inverse optimization, where the resulting efficiency curve is implied by the behavior of the reservoir manager as well as the structure of the optimization model we assume that the manager follows. The model aims to serve as a descriptive rather than a normative tool for the hydropower industry.

2 Model

2.1 Literature Context

We use structural estimation of a dynamic decision process as in Rust (1987). The premise is that if we observe a set of states and actions taken by an agent, we can work backwards to infer the objective function of that agent, by maximizing the likelihood of matching the observed data. By maximizing the likelihood function, the analyst can obtain an understanding of parameters hidden in the dynamic optimization model. In order to estimate the structural parameters in a stochastic dynamic programming problem, Rust used an algorithm he called the Nested Fixed Point (NFXP) algorithm. This algorithm has two parts, an outer loop that searches for the structural parameters with the maximum likelihood value, and an inner loop that solves the stochastic dynamic programming model given a value for the structural parameter. According to Su and Judd (2012) the NFXP algorithm is computationally demanding, because it iterates over all structural parameter values and then solves the underlying stochastic dynamic programming (SDP) model with high accuracy for each structural parameter value.

Hydropower planning problems are suitable to be treated as a stochastic dynamic problem, where a decision today change the reservoir levels and thereby affect future production opportunities. For hydropower plants operating in a well-functioning market, price can be treated as a stochastic variable, as has been explored by Wolfgang et al. (2009), among others.

2.2 Model

This subsection gives and overview of the approach. The model builds on previous work by Su and Judd (2012), Boger et al. (2017), and Brelin and Lien (2017).

A parametric approach is used for the transition probabilities between the states in the structural estimation model, as suggested by Boger et al. (2017). A state is characterized by the current electricity price, the inflow, the reservoir level, the deviation from cumulative inflow, and the deviation from the aggregate reservoir level. The parametric approach involves time series modeling of the state variables as Markovian processes, where the next state is only dependent on the previous one. The goal is to capture the dynamics of these state variables in order to successfully apply the structural estimation model.

We use weekly time steps. Inflow to the local reservoir is captured via an autoregressive process that adds a deterministic seasonal term to a residual AR(1) term. Parameters are fitted from a time series of inflow 1993–2014 from a Norwegian hydropower plant. Local inflow is one of five exogenous state variables; local reservoir level is a sixth, endogenous state variable. The other four exogenous variables are accumulated local inflow deviation, national reservoir deviation (i.e., deviation from its 10-year normal, capturing dry and wet year dynamics), spot price and forward price. We let cumulative local inflow affect the national reservoir deviation, and the national reservoir deviation affect the spot price. In this way, there will be a negative relation between local inflow and price; if there is little local inflow over a few months, then it is likely that this drought is affecting neighboring reservoirs and might be a sign of a national drought. This will likely drive prices up, and conversely.

Accumulated local inflow is specified as an exponentially weighted moving average. Accumulated local inflow deviation, which is the state variable, is the accumulation's relative deviation from its 10 year normal. All details can be found in Brelin and Lien (2017).

National reservoir deviation is an important state variable to market analysts. This information is publicly available, published weekly by the Statistics Norway, and we measure it as its relative deviation from the 10 year normal. We estimate an autoregressive process of order 1 (ARX(1)), with the accumulated local inflow deviation as an exogenous explanatory variable. In this way we capture the relationship between local inflow and regional resource state, which is an important link in the relationship between price and local inflow.

We assume that the reservoir operator for which we have data is a price taker. For the Nordic electricity market, this is reasonable. Nevertheless, we remark that since hydropower has a large share in the generation mix, future work should examine if spot prices can be regarded as exogenous to individual reservoir decisions as we do here, or whether it is necessary to take the simultaneity of storage decisions and present and future demand and supply into account.

We use the local zonal price, where the seasonality of log prices is captured by a third order Fourier series. The residual price component is an ARX(1) process, where the national reservoir deviation is the external explanatory factor. In dry years, the national reservoir deviation will be negative, and prices will be high. In wet years, the national reservoir deviation will be positive, and prices will tend to be low.

The parameters of the state dynamics above can in principle be determined not (only) by using time series of inflow, prices and national reservoir levels, but also by the observed release decisions of the reservoir operator. That is, it could be possible to try to imply the expectations that our reservoir operator has regarding e.g. inflow dynamics. Instead, we follow Rust (1987) and determine these parameters from time series only, and not release decisions. See Boger et al. (2017) for an approach to backing out price expectations.

The overall set of relationships for the state variables is a system of AR(1) equations. Included here is the reservoir balance equation, stating that the reservoir level this week equals the level last week, plus inflow, minus release and spill.

Reservoir operators are assumed to time releases from the reservoir as a trade-off between the benefit associated with immediate generation of electricity versus saving the water for future release. The immediate benefit is modeled as the spot price times within-week generation.

In order to arrive at stationarity for the Markov decision process, we follow Foss and Høst (2011), who note that since the seasonal part stays the same for different years, it will be sufficient to let the problem be conditional on week of the year. If we use time of the year as a state variable, the problem is reduced from a non-stationary to an approximately stationary problem.

As in Rust (1987), the reservoir operator has more information than the outside researchers, and we capture this by adding a payoff shock to the immediate benefit. We assume that the decision maker observes this shock, but that the researchers cannot observe it (ever). The total value of choosing a release level consists of the immediate benefit, the payoff shock, and discounted expected future benefits. This gives rise to a Bellman-like contraction that we discretize and represent as nonlinear constraints in our estimation problem, following Su and Judd (2012). The objective function in the estimation problem is maximization of the log likelihood of observing releases given our behavioral model. Payoff shocks are assumed to follow a Gumbel distribution, which allows the value function and the choice probability to be represented in closed form.

In order to discretize the Bellman contraction, we discretize the state space as well as the random errors, assumed Gaussian, that affect inflow, national reservoir deviation and spot price.

Within-week generation depends linearly on the efficiency function $E()$, which is specified to capture the power operator's resistance to deviating from the best efficiency point (BEP). The efficiency function is dependent on three factors: the BEP, ξ, the efficiency for production levels beneath the BEP, θ_1, and the efficiency for production levels above the BEP, θ_2.

For production levels (coded in the parameter d_t) below the BEP, $d_t < \xi$, the following equation applies:

$$E(\theta_1, \xi) = 1 - (\xi - d_t)\theta_1 \qquad (1)$$

and for production levels above the BEP, the efficiency function is:

$$E(\theta_2, \xi) = 1 - (d_t - \xi)\theta_2 \qquad (2)$$

Alternatives tested are a square root and a quadratic function, however, the maximum likelihood was highest for the linear form above. It is easiest to interpret our results if the power station has only one turbine-generator. We can estimate an efficiency function for a power plant with more than one unit, or even several power plants, however, it becomes difficult to compare with the real efficiency curves if they are available.

3   Case Study and Results

We consider a hydropower plant located in the Norwegian electricity price zone NO4. The model was implemented in AMPL (A Mathematical Programming Language), (Fourer et al. 1990). The environment that was used was the AMPL IDE, and the solver used was the Artelys KNITRO solver (Byrd et al. 2006). The KNITRO solver is used in large-scale nonlinear optimization, and it was not limited in memory usage. The computers had 32 GB of memory installed, which allowed us to discretize the production sufficiently so that we could capture the differences between efficient and non-efficient production levels. The model had a total of 25442 variables and 25440 constraints. The two extra variables are θ_1 and θ_2. The CPU was an Intel Core i7-6700 with a maximum speed of 3.40 GHz. Solving the model for a manually set value of θ_1 and θ_2 takes approximately 40 s. Alternatively, solving the model with variable values for θ_1 and θ_2 will take between 5 and 12 h.

The results indicate that the reservoir managers require a 51% higher reward for producing at 100% instead of 83% of maximum production (83% is the BEP). They require a 17% higher reward for producing at 67% of maximum production, i.e. below BEP. Further, since the relationship is assumed to be linear, they require a $2 \cdot 17\% = 34\%$ higher reward for producing at 50% of maximum production, and so on.

The implied efficiency loss of producing above the BEP is clearly larger than the physical loss when producing below the BEP. Since the efficiency of the turbine-generator scheme does not lose as much as 51% when producing at maximum, there has to be some other considerations behind the producer's choice. This means that the reservoir manager appears reluctant to increase weekly generation beyond the BEP. One issue is that it is not normal to produce at full capacity over an entire week, due to the extra wear such a pattern induces. We think this is a major explanatory factor. Other interesting explanations include irrationality (see Alnæs et al. (2015)), or market power.

A plot of the implied penalty is shown in Fig. 1, together with the efficiency curve of a Francis turbine. The efficiency curve of a Francis turbine is included in the plot, as this is the most used turbine for hydropower plants that share the same technical specifications as the power plant. We have discretized the production levels into 6, where level 1 is the minimum and level 6 the maximum, the latter corresponding to maximum generation. The y-axes shows the efficiency in percent. Comparing the two curves, it is important to keep in mind that the implied curve comes from a model with weekly resolution. It is reassuring to see that the maximum implied efficiency is indeed at the BEP for the Francis turbine, at 83%.

It is important to keep in mind that the discovered efficiency curves represent the behavior of the reservoir managers, given that the behavioral (optimization) model is correct. The implied efficiency does not have to match the mechanical efficiency of the turbine-generator scheme but can include other hidden economic factors that explain the dissimilarities between the two curves. There can be several reasons why a power producer does not want to produce at other production levels. Examples of these are increased maintenance cost, lowered durability, and cavitation problems.

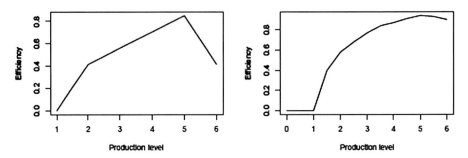

Fig. 1. Comparison between the implied efficiency and the efficiency of a Francis turbine.

According to Kumar and Saini (2010), the point of critical cavitation is above the BEP. The important implication of this is that producing at a higher level than the BEP leads to a much higher cavitation than producing at or below the BEP. As such, it is obviously not just the pure efficiency loss that is being taken into account by the decision-makers at the plant. Damage to the turbine becomes an important factor when deciding whether to produce above the BEP. This may explain why the reservoir managers require such a high reward for doing so. Damaging the turbine can be costly, and as a result, the compensation must be high.

The results of the implied efficiency seen in Fig. 1 can be the basis for internal discussions in the hydropower company as to whether or not the required extra reward reflects their operating policy and if their current policy is reasonable. The implied efficiency should be the result of potential mechanical failure and related costs, the production efficiency and the mechanical fatigue this production implies. Ultimately, the power producer wants to produce power at levels and at times which generate the highest total profits.

The data series used to estimate the θs for the power producer contains over 20 years of production data. To examine if the plant's preferences have changed over time, the data series was split in half and the model was solved for the first and second halves with the results displayed in Table 1. The goal was to identify if the willingness to deviate from the efficient production level has changed during this time.

Table 1. Changing values of θ_1 and θ_2 for the power plant over time

	First half of the sample (20 years)	Second half
θ_1	0.5456	0.0794
θ_2	1	0.398
Log-likelihood	−579.69	−676.54

The results displayed in Table 1 show that the willingness to deviate from the BEP has increased over the years, as both θs are lower for the second half than the first half. A potential explanation is the change in monitoring technology. It is possible that the power plant had a much stricter policy for production at levels above the BEP due to uncertainty regarding the potential economic downside. A θ_2 value of 1 signifies that the producer should not want to produce above the BEP, as this would mean that the producer receives no power in return for the released water. A potential scenario is that the precision of monitoring technology related to turbine maintenance and wear was increased. The results of such monitoring may have changed their willingness from being completely unwilling to produce at levels above the BEP in fear of, among other things, cavitation, to producing at levels above the BEP if the economic relative gain was above 39.8%.

4 Conclusions

A structural estimation model has been applied to data series from a hydropower plant in Norway. This has been done in order to discover the reservoir managers' preferences related to the willingness to produce at other levels of than the production level with the highest efficiency. The results yield an implied efficiency curve that considers the producer's preferences instead of the mechanical efficiency. The implied efficiency is a valuable finding and can be used as a basis for internal discussion within the hydropower plant's management.

The results show a greater willingness of the manager to produce at levels below than above the best efficiency point. Our hypothesis is that this is mainly linked to cavitation issues when producing at higher levels. Changes in these preferences over time was also investigated, which showed an increased willingness to produce both above and below the best efficiency point at the end of the time period.

References

Alnæs, E.N., Grøndahl, R., Boomsma, T.K., Fleten, S.-E.: Insights from actual day-ahead bidding of hydropower. Int. J. Sustain. Energy Plan. Manag. **7**, 34–54 (2015)

Brelin, S., Lien, M.A.: Empirical analysis of hydropower scheduling. Master thesis, Norwegian University of Science and Technology (2017)

Boger, M., Fleten, S.-E., Pichler, A., Keppo, J., Vestbøstad, E.M.: Backing out expectations from hydropower release time series. In: IAEE International Conference, Singapore (2017)

Byrd, R.H., Nocedal, J., Waltz, R.A.: KNITRO: an integrated package for nonlinear optimization. In: di Pillo, G., Roma, M. (eds.) Large-Scale Nonlinear Optimization, pp. 35–59. Springer, New York (2006)

Foss, M.Ø., Høst, A.: Hydroelectric Real Options: A Structural Estimation Approach. Masters thesis, Norwegian University of Science and Technology (2011)

Fourer, R., Gay, D.M., Kernighan, B.W.: A modeling language for mathematical programming. Manag. Sci. **36**(5), 519–554 (1990)

Kumar, P., Saini, R.: Study of caviation in hydro turbines - a review. Renew. Sustain. Energy Rev. **14**, 374–383 (2010)

Rust, J.: Optimal replacement of GMC bus engines: an empirical model of Harold Zurcher. Econometrica **55**(5), 999–1033 (1987)

Su, C.L., Judd, K.L.: Constrained optimization approaches to estimation of structural models. Econometrica **80**(5), 2213–2230 (2012)

Wolfgang, O., Haugstad, A., Mo, B., Gjelsvik, A., Wangsteen, I., Doorman, G.: Hydro reservoir handling in Norway before and after deregulation. Energy **34**(10), 1642–1651 (2009)

Norway as a Battery for the Future European Power System – Comparison of Two Different Methodological Approaches

Ingeborg Graabak[1(✉)], Stefan Jaehnert[1], Magnus Korpås[2], and Birger Mo[1]

[1] SINTEF Energy Research, Trondheim, Norway
Ingeborg.Graabak@sintef.no
[2] Norwegian University of Science and Technology (NTNU), Trondheim, Norway

Abstract. This paper compares the simulation results for two stochastic optimization power market models. EMPS uses aggregation and heuristics to calculate the optimal dispatch. SOVN simulates the operation of the power system in one large linear programming problem taking each single plant and reservoir into consideration. The comparison is for a future system in Europe where wind and solar power production supplies 61% of the annual demand. Three different alternatives for the Norwegian hydropower system is studied: present generation capacity (about 30 GW), increased capacity to about 41 GW and further to about 49 GW. The analyses show that SOVN to a larger degree than EMPS manage to increase production in high price periods and pump in low price periods. This particularly applies to the weeks before the change from the depletion (winter) to the filling (summer) period. This better ability to exploit the flexibility of the hydropower system is due to applying a formal optimization in SOVN compared to advanced heuristics in EMPS. For regions without pumping possibility, there is less difference between the models.

Keywords: Stochastic power market optimisation models
Increases in hydropower capacities · Pumped storage

1 Introduction

The future European power system is expected to include large shares of variable wind and solar power resources. Reference [1] shows that Norwegian hydropower can balance part of the variability and significantly decrease peak and average power prices in neighbouring countries like UK, Germany, the Netherlands and France in 2050. The reference shows results from analyses with two stochastic optimisation models, EMPS and SOVN. Due to the application of a formal optimisation in SOVN compared to heuristics in EMPS, the hydropower system flexibility can be exploited much better. Hence, analysed with SOVN the power prices decrease more than analysed with EMPS. While [1] analyses impacts on power prices, this paper compares results from the two models mainly related to power production, and development of energy content in reservoirs for the power system of Northern Europe in 2050.

© Springer Nature Switzerland AG 2019
A. Helseth (Ed.): HSCM 2018, *Proceedings of the 6th International Workshop on Hydro Scheduling in Competitive Electricity Markets*, pp. 76–83, 2019.
https://doi.org/10.1007/978-3-030-03311-8_10

2 Objective

The objective of this paper is to compare the results from two stochastic dynamic optimization models with different methodological approaches for the simulation of the power system. Previous research compared the models for the Nordic region in 2020 [2]. This paper expands the analysis to Europe in 2050 and a power system with very high shares of wind and solar resources in the production portfolio.

3 Methodology

3.1 Models

A potential future power system in Europe is analysed by two stochastic optimisation and power market simulators, EMPS [3] and SOVN [2]. Both models maximize the expected total economic surplus in the simulated system through the dispatch of generation, given a consumption profile and transmission constraints.

One of the EMPS' strengths is an advanced representation of future cost of power systems operation with energy storage. There is no significant production cost for hydropower. However, with stochastic inflow and limited hydro storage determination of an optimal strategy for hydropower generation becomes a complex problem. EMPS executes two phases: the strategy and the simulation phase. In the first phase, water values for each reservoir are calculated as option values of the stored energy for different operational strategies. In the second phase, the operation of the power system is optimized and simulated for the different stochastic outcomes (climatic years). The model optimizes the power dispatch in each time step per node. The optimization procedure starts with calculating the optimal dispatch with hydropower aggregated to one plant and one reservoir per node/region. In a next step, the aggregated production is distributed on the individual hydropower plants based on advanced heuristics. This ruled-based procedure verifies if the desired production at aggregated level is obtainable within all constraints at the detailed level. If the aggregated production is not possible taking all details in the hydropower system into consideration, the loop continues with a new dispatch at aggregated level and a new reservoir drawdown procedure etc.

In contrary to the mixture of optimization and heuristic in EMPS, the SOVN model uses a formal optimisation, when determining the dispatch of the individual hydropower plants in the detailed water courses. The drawback is very long calculation time. The way the SOVN model is used in the following analysis, the difference between SOVN and EMPS is the optimisation of the hydropower utilization in the operational problem only. Except for this difference, the models are run in an equal way. The reason for using the same strategy for SOVN and EMPS is that it requires weeks to calculate water values by the SOVN prototype model. Two aspects with the optimisation of the hydropower utilization are particularly important for the results:

(i) SOVN has a better representation of short-term flexibility e.g. pumped-storage. EMPS seldom pumps in the winter due to its rule-based heuristics methods.

(ii) SOVN distributes the water in the long cascade coupled rivers system such that plants with high capacity have as much water as possible upstream to the plant. EMPS distribute the water such that the risk for empty reservoirs in the winter or overflow in the spring and summer is minimised.

The consequence is that SOVN has more water available for production in high price periods particularly in the late winter/spring [1].

The main inputs to both models include costs and generation capacities, net transmission capacities and electricity consumption with price elasticity and information about historical climate variables like temperatures, hydro inflow, wind, solar radiation, typically with hourly resolution. The output from the models is a detailed dispatch of the power system, including among other power balances, exchange and prices.

This study has 59 nodes for the whole Europe. Each node has an endogenously determined internal supply and demand balance with distinct import and export transmission capacities to the neighbouring nodes. Norway, Sweden, UK and Germany are modelled with 11, 6, 6 and 7 regions respectively, while other countries in Europe are more aggregated modelled. Reference [4] shows a full European map with all the regions. Figure 1 shows the regions that are focused in these EMPS and SOVN analyses.

Hydropower in the Nordic area is described as detailed water courses with multiple power plants in series or parallel. The description includes minimum and maximum reservoir levels, minimum discharge requirements and others. The remaining European countries use an aggregated model for the hydropower.

The temporal resolution of both models is flexible, but calculation time increases significantly with more time steps. This present analysis uses 2 h resolution for weekdays and 4 h for weekends. To remain with computational feasibility, the possibility to include start- up costs for thermal power plants could not be applied.

3.2 Scenario Data

The EU 7th Framework project eHighway2050 scenario X-7 is used for quantification of the future European power system [5]: generation capacities per region, demand, transmission capacities between regions and fuel prices. Figure 1 to the right shows the annual power production per generation type for the whole Europe for the X-7 scenario. The annual consumption aggregated for Europe is 4277 TWh. Wind and solar resources supply about 61% of the demand in the scenario. Wind and solar resources are Reanalysis data [6] for the period 1948 to 2005. Reference [7] describes the modelling of wind and solar data.

The hydropower reservoirs in Norway represent approximately half of the total hydro storage capacity in Europe with about 85 TWh of storage [8]. The main purpose of hydro reservoirs in Norway is to store water from the warm season (summer) to the cold season (winter) and from wet years to dry years. The Norwegian consumption has a strong seasonal profile due to electric heating. Furthermore, the inflow to the system is limited in the winter due to precipitation in terms of snow. The reservoirs are filled up in the spring and in the summer because of snow melting and rainy periods. Modelling of the Nordic hydropower system is from the EU 7th Framework project TWENTIES. The present production capacity in the Norwegian power system is about 30 GW. Previous research shows possibilities for increases of capacities in existing hydropower plants in Southern Norway [9]. The capacity is increased from its present value ca 41 GW (11 GW extra production capacity) and further to ca 49 GW (19 GW extra production capacity) respectively. The inflow to the Nordic hydropower system is represented by 75 years of historical data. Table 1 shows the increases distributed on four EMPS/SOVN regions in southern Norway (see Fig. 1).

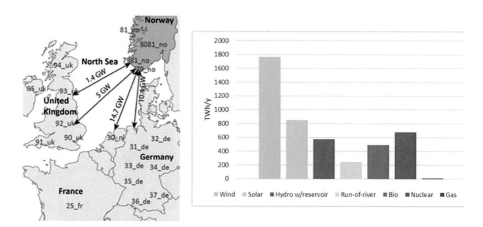

Fig. 1. To the left: EMPS and SOVN regions mainly focused in these analyses. To the right: yearly power production in Europe per technology eHighway2050 X-7 scenario, EMPS analysis.

Table 1. Increases in hydro generation capacities in four Norwegian regions

EMPS/SOVN region (see Fig. 1)	Present capacity [GW]	New capacity 11 GW [GW]	Pump capacity 11 GW [GW]	New capacity 19 GW [GW]	Pump capacity 19 GW [GW]
79_no	4.1	7.6	1.4	8.3	1.4
7981_no	3.6	7.8	2.1	10.1	3.4
81_no	5	7.9	0	8.5	0
8081_no	2.1	3.1	1	6.3	4.4
Total	14.8	26.4	4.5	33.2	9.2

4 Results

The paper compares EMPS and SOVN results related to: (i) Production at aggregated level (region) and for a single plant and (ii) Reservoir handling at aggregated level. Figure 2 shows net hydropower production in two regions, 79_no and 81_no, averaged hour-by-hour for 75 climatic years. The regions 7981_no and 8081_no have similar patterns as 79_no. For these three regions, the resulting production patterns are significant different for EMPS and SOVN. SOVN pumps much more than EMPS in periods with low prices. Due to the pumping, there is more energy available for production in high price periods. Furthermore, as shown in Fig. 2, there are small differences between EMPS and SOVN production patterns for region 81_no, as there is no pumping capacity in the region (see Table 1). Thus, there is no extra flexibility. However, for region 81_no we observe that SOVN produces more in the winter and less in the summer than EMPS. EMPS has less production in the late winter due to its seasonal heuristic approach.

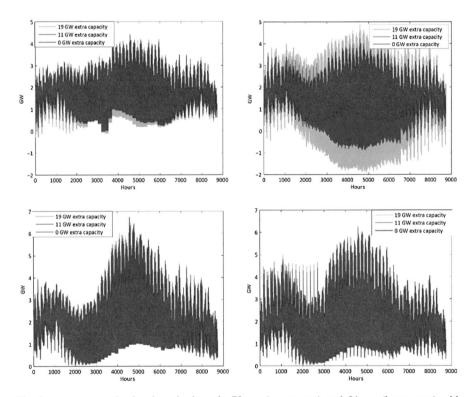

Fig. 2. Average production hour-by-hour in 79_no (upper row) and 81_no (lower row) with increased hydropower capacities for 75 simulation years, EMPS (left) and SOVN (right) results

Figure 3 shows the average prices in 79_no corresponding to the production shown in the upper part of Fig. 2. Comparison of the two figures shows that SOVN manage to increase production in high price periods and thereby smooth out prices to a much larger degree than EMPS.

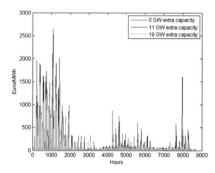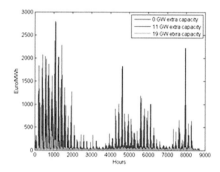

Fig. 3. Average prices hour-by-hour in 79_no for 75 years with simulations, EMPS (left) and SOVN (right)

Figure 4 shows the hydropower production for the Kvilldal plant simulated by EMPS and by SOVN. The results are similar for other plants with pumping capacity. As in mentioned in Sect. 3.1, one important difference between SOVN and EMPS, is that EMPS hardly pumps in the winter months. This can also be observed in Fig. 4. In the winter months, there is limited inflow to the reservoirs. According to the heuristic in EMPS, the reservoirs are depleted such that the relative water values for reservoirs in the same river system are approximately the same. With limited difference in water values between the reservoirs, there will not be any pumping. However, in the summer, when there is significant inflow to the reservoirs but only minor production, pumping is used in EMPS to avoid spillage from the reservoirs.

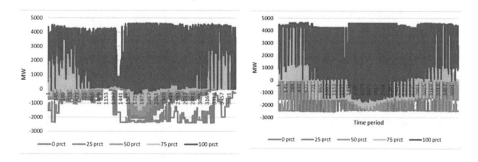

Fig. 4. Percentiles for hydropower production for Kvilldal plant in region 7981_no, time period-by-time period for 75 simulation years, EMPS results to the left and SOVN results to the right Generation capacity increased to 4.6 GW (19 GW increase in Norway). New pump capacity with 2.4 GW. Capacity of upstream reservoir: 237 MM3.

Figure 5 shows the 0, 50 and 100 percentiles for development of the energy content in reservoirs for EMPS and SOVN for the regions 79_no and 81_no. EMPS and SOVN use different methodologies for optimisation of the use of the water but the energy content in the reservoirs is calculated in the same way.

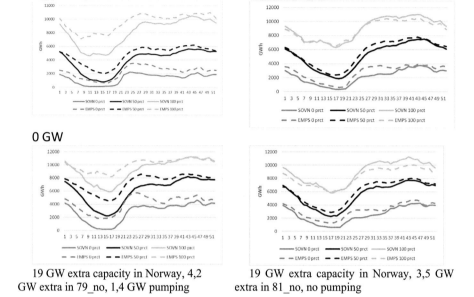

19 GW extra capacity in Norway, 4,2 GW extra in 79_no, 1,4 GW pumping

19 GW extra capacity in Norway, 3,5 GW extra in 81_no, no pumping

Fig. 5. Percentiles for reservoir development region 79_no and 81_no week-by-week for 75 simulations years, EMPS and SOVN analysis

For both regions we observe that the energy content in the aggregated reservoir increase with increased hydro generation capacity. With increased capacity, there will be less probability for overflow. Thus, more water can be stored in the reservoirs and the energy content increases. It increases more for 79_no which has pump capacity and can pump water to higher reservoirs in low-price periods.

For 79_no the largest difference in reservoir level between EMPS and SOVN is around the weeks where EMPS change seasonal strategy and goes from depletion (winter) to filling (summer). In the last weeks of the depletion period, EMPS will have limited water left in reservoirs upstream to plants with increased capacity. All reservoirs are depleted in such a way that they have about the same risk of spillage in the coming spring (melting) inflow period. SOVN will, if possible, distribute water between reservoirs in such a way that there is water available upstream to plants with increased capacity. Thus, SOVN can produce more in the weeks before the melting starts (about week 17), and the energy content in the reservoirs will be lower. As mentioned in Sect. 3.1, SOVN uses in these analysis water values from EMPS. These values are too low for SOVN for the 79_no 0 GW case. The reservoirs are empty in long periods (the 0 percentile). Region 81_no does not have any pumping capacity. For this region, there is less difference in development of energy content between EMPS/SOVN, than for 79_no with pumping capacity.

5 Conclusions

This study compares analysis results from two different stochastic optimization models: EMPS and SOVN. The analyses show that SOVN to a larger degree than EMPS manage to increase production in high price periods and pumping in low price periods. This particularly applies to the weeks before the change from the depletion to the filling period. This better ability to exploit the flexibility of the hydropower system is due to applying a formal optimization in SOVN compared to advanced heuristics in EMPS. Power production particularly increases in high price periods with SOVN compared to EMPS for regions with pumping capacity. For a region without pumping capacity there is less differences between the models.

References

1. Graabak, I., Korpaas, M., Jaehnert, S., Belsnes, M.: Balancing future variable wind and solar power production in Northern Europe with Norwegian hydro power (2017, submitted)
2. Helseth, A., Mo, B., Henden, A., Warland, G.: Detailed long-term hydro-thermal scheduling for expansion planning in the Nordic power system. IET Res. J. (2017). ISSN 1751-8644
3. Wolfgang, O., Haugstad, A., Mo, B., Gjelsvik, A., Wangensteen, I., Doorman, G.: Hydro reservoir handling in Norway before and after deregulation. Energy **34**, 1642–1651 (2009)
4. Graabak, I., Jaehnert, S., Korpås, M., Mo, B.: Norway as a battery for the future european power system - impacts on the hydropower system. Energies **10**, 2054 (2017). Special Issue. Hydropower 2017
5. Bruninx, K., et al.: D 2.1 Data sets of scenarios for 2050 (2015). http://www.e-highway2050.eu/results/. Accessed June 2017
6. Kalnay, E., Kanamitsu, M., Kistler, R., Collins, W., Deaven, D., Gandin, L., et al.: The NCEP/NCAR 40-year reanalysis project. Bull. Am. Soc. **77**, 437–470 (1996)
7. Svendsen, H.: Hourly wind and solar energy time series from Reanalysis dataset. SINTEF Energy Research. Project Memo 2017. http://hdl.handle.net/11250/2468143. Accessed June 2017
8. Eurelectric: Hydropower for a sustainable Europe (2013). http://www.eurelectric.org/media/26690/hydro_report_final-2011-160-0011-01-e.pdf. Accessed June 2017
9. Solvang, E., Harby, A., Killingtveit, Å.: Increased balancing power capacity in Norwegian hydroelectric power stations. SINTEF Energy Research, TR A7195 (2012). ISBN 9788259435194

Author Index

© Springer Nature Switzerland AG 2019
A. Helseth (Ed.): HSCM 2018, *Proceedings of the 6th International Workshop on Hydro Scheduling in Competitive Electricity Markets*, p. 85, 2019.
https://doi.org/10.1007/978-3-030-03311-8